Simon Newcomb

Elements of Differential and Integral Calculus

Simon Newcomb

Elements of Differential and Integral Calculus

ISBN/EAN: 9783337811402

Printed in Europe, USA, Canada, Australia, Japan

Cover: Foto ©berggeist007 / pixelio.de

More available books at **www.hansebooks.com**

NEWCOMB'S MATHEMATICAL COURSE

ELEMENTS

OF THE

DIFFERENTIAL AND INTEGRAL

CALCULUS.

BY

SIMON NEWCOMB

Professor of Mathematics in the Johns Hopkins University

NEW YORK
HENRY HOLT AND COMPANY
1887

COPYRIGHT, 1887,

BY

HENRY HOLT & CO.

PREFACE.

THE present work is intended to contain about as much of the Calculus as an undergraduate student, either in Arts or Science, can be expected to master during his regular course. He may find more exercises than he has time to work out; in this case it is suggested that he only work enough to show that he understands the principles they are designed to elucidate.

The most difficult question which arises in treating the subject is how the first principles should be presented to the mind of the beginner. The author has deemed it best to begin by laying down the logical basis on which the whole superstructure must ultimately rest. It is now well understood that the method of limits forms the only rigorous basis for the infinitesimal calculus, and that infinitesimals can be used with logical rigor only when based on this method, that is, when considered as quantities approaching zero as their limit. When thus defined, no logical difficulty arises in their use; they flow naturally from the conception of limits, and they are therefore introduced at an early stage in the present work.

The fundamental principles on which the use of infinitesimals is based are laid down in the second chapter. But it is not to be expected that a beginner will fully grasp these principles until he has become familiar with the mechanical process of differentiation, and with the application of the calculus to special problems. It may therefore be found best to begin with a single careful reading of the chapter, and afterward to use it for reference as the student finds occasion to apply the principles laid down in it.

The author is indebted to several friends for advice and assistance in the final revision of the work. Professor John E. Clark of the Sheffield Scientific School and Dr. Fabian Franklin of the Johns Hopkins University supplied suggestions and criticisms which proved very helpful in putting the first three chapters into shape. Miss E. P. Brown of Washington has read all the proofs, solving most of the problems as she went along in order to test their suitability.

CONTENTS.

PART I.

THE DIFFERENTIAL CALCULUS.

                                                                          PAGE

CHAPTER I. OF VARIABLES AND FUNCTIONS.................. 3
§ 1. Nature of Functions. 2. Their Classification. 3. Functional Notation. 4. Functions of Several Variables. 5. Functions of Functions. 6. Product of the First n Numbers. 7. Binomial Coefficients. 8. Graphic Representation of Functions. 9. Continuity and Discontinuity of Functions. 10. Many-valued Functions.

CHAPTER II. OF LIMITS AND INFINITESIMALS................ 17
§ 11. Limits. 12. Infinites and Infinitesimals. 13. Properties. 14. Orders of Infinitesimals. 15. Orders of Infinites.

CHAPTER III. OF DIFFERENTIALS AND DERIVATIVES.......... 25
§ 16. Increments of Variables. 17. First Idea of Differentials and Derivatives. 18. Illustrations. 19. Illustration by Velocities. 20. Geometrical Illustration.

CHAPTER IV. DIFFERENTIATION OF EXPLICIT FUNCTIONS...... 31
§ 21. The Process of Differentiation in General. 22. Differentials of Sums. 23. Differential of a Multiple. 24. Differential of a Constant. 25. Differentials of Products and Powers. 26. Differential of a Quotient of Two Variables. 27. Differentials of Irrational Expressions. 28. Logarithmic Functions. 29. Exponential Functions. 30. The Trigonometric Functions. 31. Circular Functions. 32. Logarithmic Differentiation. 33. Velocity or Derivative with Respect to the Time.

CHAPTER V. FUNCTIONS OF SEVERAL VARIABLES AND IMPLICIT FUNCTIONS....................................... 54
§ 34. Partial Differentials and Derivatives. 35. Total Differentials. 36. Principles involved in Partial Differentiation. 37. Dif-

lus to special problems. It may therefore be found best to begin with a single careful reading of the chapter, and afterward to use it for reference as the student finds occasion to apply the principles laid down in it.

The author is indebted to several friends for advice and assistance in the final revision of the work. Professor John E. Clark of the Sheffield Scientific School and Dr. Fabian Franklin of the Johns Hopkins University supplied suggestions and criticisms which proved very helpful in putting the first three chapters into shape. Miss E. P. Brown of Washington has read all the proofs, solving most of the problems as she went along in order to test their suitability.

CONTENTS.

PART I.

THE DIFFERENTIAL CALCULUS.

                                                                            PAGE

CHAPTER I. OF VARIABLES AND FUNCTIONS.................. 3
§ 1. Nature of Functions. 2. Their Classification. 3. Functional Notation. 4. Functions of Several Variables. 5. Functions of Functions. 6. Product of the First n Numbers. 7. Binomial Coefficients. 8. Graphic Representation of Functions. 9. Continuity and Discontinuity of Functions. 10. Many-valued Functions.

CHAPTER II. OF LIMITS AND INFINITESIMALS................ 17
§ 11. Limits. 12. Infinites and Infinitesimals. 13. Properties. 14. Orders of Infinitesimals. 15. Orders of Infinites.

CHAPTER III. OF DIFFERENTIALS AND DERIVATIVES.......... 25
§ 16. Increments of Variables. 17. First Idea of Differentials and Derivatives. 18. Illustrations. 19. Illustration by Velocities. 20. Geometrical Illustration.

CHAPTER IV. DIFFERENTIATION OF EXPLICIT FUNCTIONS...... 31
§ 21. The Process of Differentiation in General. 22. Differentials of Sums. 23. Differential of a Multiple. 24. Differential of a Constant. 25. Differentials of Products and Powers. 26. Differential of a Quotient of Two Variables. 27. Differentials of Irrational Expressions. 28. Logarithmic Functions. 29. Exponential Functions. 30. The Trigonometric Functions. 31. Circular Functions. 32. Logarithmic Differentiation. 33. Velocity or Derivative with Respect to the Time.

CHAPTER V. FUNCTIONS OF SEVERAL VARIABLES AND IMPLICIT FUNCTIONS...................................... 54
§ 34. Partial Differentials and Derivatives. 35. Total Differentials. 36. Principles involved in Partial Differentiation. 37. Dif-

ferentiation of Implicit Functions. 38. Implicit Functions of Several Variables. 39. Case of Implicit Functions expressed by Simultaneous Equations. 40. Functions of Functions. 41. Functions of Variables, some of which are Functions of the Others. 42. Extension of the Principle. 43. Nomenclature of Partial Derivatives. 44. Dependence of the Derivative upon the Form of the Function.

CHAPTER VI. DERIVATIVES OF HIGHER ORDERS.............. 74
§ 45. Second Derivatives. 46. Derivatives of Any Order. 47. Special Forms of Derivatives of Circular and Exponential Functions. 48. Successive Derivatives of an Implicit Function. 49. Successive Derivatives of a Product. 50. Successive Derivatives with Respect to Several Equicrescent Variables. 51. Result of Successive Differentiations independent of the Order of the Differentiations. 52. Notation for Powers of a Differential or Derivative.

CHAPTER VII. SPECIAL CASES OF SUCCESSIVE DERIVATIVES... 86
§ 53. Successive Derivatives of a Power of a Derivative. 54. Derivatives of Functions of Functions. 55. Change of the Equicrescent Variable. 56. Two Variables connected by a Third.

CHAPTER VIII. DEVELOPMENTS IN SERIES................... 95
§ 57. Classification of Series. 58. Convergence and Divergence of Series. 59. Maclaurin's Theorem. 60. Ratio of the Circumference of a Circle to its Diameter. 61. Use of Symbolic Notation for Derivatives. 62. Taylor's Theorem. 63. Identity of Taylor's and Maclaurin's Theorems. 64. Cases of Failure of Taylor's and Maclaurin's Theorems. 65. Extension of Taylor's Theorem to Functions of Several Variables. 66. Hyperbolic Functions.

CHAPTER IX. MAXIMA AND MINIMA OF FUNCTIONS OF A SINGLE VARIABLE... 117
§ 67. Definition of Maximum Value and Minimum Value. 68. Method of finding Maximum and Minimum Values of a Function. 69. Case when the Function which is to be a Maximum or Minimum is expressed as a Function of Two or More Variables connected by Equations of Condition.

CHAPTER X. INDETERMINATE FORMS......................... 128
§ 70. Examples of Indeterminate Forms. 71. Evaluation of the Form $\frac{0}{0}$. 72. Forms $\frac{\infty}{\infty}$ and $0 \times \infty$. 73. Form $\infty - \infty$. 75. Forms $0°$ and $\infty°$.

CONTENTS.

CHAPTER XI. OF PLANE CURVES.................... 137
§ 76. Forms of the Equations of Curves. 77. Infinitesimal Elements of Curves. 78. Properties of Infinitesimal Arcs and Chords. 79. Expressions for Elements of Curves. 80. Equations of Certain Noteworthy Curves. The Cycloid. 81. The Lemniscate. 82. The Archimedean Spiral. 83. The Logarithmic Spiral.

CHAPTER XII. TANGENTS AND NORMALS................ 147
§ 84. Tangent and Normal compared with Subtangent and Subnormal. 85. General Equation for a Tangent. 86. Subtangent and Subnormal. 87. Modified Forms of the Equation. 88. Tangents and Normals to the Conic Sections. 89. Length of the Perpendicular from the Origin upon a Tangent or Normal. 90. Tangent and Normal in Polar Co-ordinates. 91. Perpendicular from the Pole upon the Tangent or Normal. 92. Equation of Tangent and Normal derived from Polar Equation of the Curve.

CHAPTER XIII. OF ASYMPTOTES, SINGULAR POINTS AND CURVE-TRACING 157
§ 93. Asymptotes. 94. Examples of Asymptotes. 95. Points of Inflection. 96. Singular Points of Curves. 97. Condition of Singular Points. 98. Examples of Double-points. 99. Curve-tracing.

CHAPTER XIV. THEORY OF ENVELOPES.................. 169
§ 100. Envelope of a Family of Lines. 101. All Lines of a Family tangent to the Envelope. 102. Examples and Applications.

CHAPTER XV. OF CURVATURE, EVOLUTES AND INVOLUTES..... 180
§ 103. Position; Direction; Curvature. 104. Contacts of Different Orders. 105. Intersection or Non-intersection of Curves according to the Order of Contact. 106. Radius of Curvature. 107. The Osculating Circle. 108. Radius of Curvature when the Abscissa is not taken as the Independent Variable. 109. Radius of Curvature of a Curve referred to Polar Co-ordinates. 110. Evolutes and Involutes. 111. Case of an Auxiliary Variable. 112. The Evolute of the Parabola. 113. Evolute of the Ellipse. 114. Evolute of the Cycloid. 115. Fundamental Properties of the Evolute. 116. Involutes.

PART II.

THE INTEGRAL CALCULUS.

CHAPTER I. THE ELEMENTARY FORMS OF INTEGRATION....... 201
§ 117. Definition of Integration. 118. Arbitrary Constant of Integration. 119. Integration of Entire Functions. 120. The Logarithmic Function. 121. Another Method of obtaining the Logarithmic Integral. 122. Exponential Functions. 123. The Elementary Forms of Integration.

CHAPTER II. INTEGRALS IMMEDIATELY REDUCIBLE TO THE ELEMENTARY FORMS.. 209
§ 124. Integrals reducible to the Form $\int y^n dy$. 125. Application to the Case of a Falling Body. 126. Reduction to the Logarithmic Form. 127. Trigonometric Forms. 128. Integration of $\frac{dx}{a^2 + x^2}$ and $\frac{dx}{a^2 - x^2}$. 129. Integrals of the Form $\int \frac{dx}{a + bx + cx^2}$. 130. Inverse Sines and Cosines as Integrals. 131. Two Forms of Integrals expressed by Circular Functions. 132. Integration of $\frac{dx}{\sqrt{a^2 \mp x^2}}$. 133. Integration of $\frac{dx}{\sqrt{a + bx \pm cx^2}}$. 134. Exponential Forms.

CHAPTER III. INTEGRATION BY RATIONAL TRANSFORMATIONS.. 222
§ 135. Integration of $\frac{(a + x)^m}{x^n} dx$, $\frac{x^m dx}{(a + bx)^n}$ and $\frac{x dx}{a + bx \pm cx^2}$. 136. Reduction of Rational Fractions in general. 137. Integration by Parts.

CHAPTER IV. INTEGRATION OF IRRATIONAL ALGEBRAIC DIFFERENTIALS.. 233
§ 138. When Fractional Powers of the Independent Variable enter into the Expression. 139. Cases when the Given Differential Contains an Irrational Quantity of the Form $\sqrt{a + bx + cx^2}$. 140. Integration of $d\theta = \frac{dr}{r\sqrt{ar^2 + br - 1}}$. 141. General Theory of Irrational Binomial Differentials. 142. Special Cases when $m + 1 = n$, or $m + 1 + np = -n$. 143. Forms of Reduction of Irrational Binomials. 144. Formulæ A and B, in which m is increased or diminished by n. 145. Formulæ C and D, in which p is increased or diminished by 1. 146. Effect of the Formulæ. 147. Case of Failure in this Reduction.

CONTENTS.

CHAPTER V. INTEGRATION OF TRANSCENDENT FUNCTIONS..... 246

§ 148. Integration of $\int e^{mx} \cos nx\, dx$ and $\int e^{mx} \sin nx\, dx$. 149. Integration of $\sin^m x \cos^n x\, dx$. 150. Special Cases of $\int \sin^m x \cos^n x\, dx$. 151. Integration of $\dfrac{dx}{m^2 \sin^2 x + n^2 \cos^2 x}$. 152. Integration of $\dfrac{dy}{a + b \cos y}$. 153. Special Cases of the Last Two Forms. 154. Integration of $\sin mx \cos nx\, dx$. 155. Integration by Development in Series.

CHAPTER VI. OF DEFINITE INTEGRALS....................... 255

§ 156. Successive Increments of a Variable. 157. Differential of an Area. 158. The Formation of a Definite Integral. 159. Two Conceptions of a Definite Integral. 160. Differentiation of a Definite Integral with respect to its Limits. 161. Examples and Exercises in finding Definite Integrals. 162. Failure of the Method when the Function becomes Infinite. 163. Change of Variable in Definite Integrals. 164. Subdivision of a Definite Integral. 165. Definite Integrals through Integration by Parts.

CHAPTER VII. SUCCESSIVE INTEGRATION................... 272

§ 166. Differentiation under the Sign of Integration. 167. Application of the Principle to Definite Integrals. 168. Integration by means of Differentiating Known Integrals. 169. Application to a Special Case. 170. Double Integrals. 171. Value of a Function of Two Variables obtained from its Second Derivative. 172. Triple and Multiple Integrals. 173. Definite Double Integrals. 174. Definite Triple and Multiple Integrals. 175. Product of Two Definite Integrals. 176. The Definite Integral $\int_{-\infty}^{+\infty} e^{-x^2} dx$.

CHAPTER VIII. RECTIFICATION AND QUADRATURE............ 285

§ 177. The Rectification of Curves. 178. The Parabola. 179. The Ellipse. 180. The Cycloid. 181. The Archimedean Spiral. 182. The Logarithmic Spiral. 183. The Quadrature of Plane Figures. 184. The Parabola. 185. The Circle and the Ellipse. 186. The Hyperbola. 187. The Lemniscate. 188. The Cycloid.

CHAPTER IX. THE CUBATURE OF VOLUMES.................. 297

§ 189. General Formulæ. 190. The Sphere. 191. The Pyramid. 192. The Ellipsoid. 193. Volume of any Solid of Revolution. 194. The Paraboloid of Revolution. 195. The Volume generated by the Revolution of a Cycloid around its Base. 196. The Hyperboloid of Revolution of Two Nappes. 197. Ring-shaped Solids of Revolution. 198. Application to the Circular Ring. 199. Quadrature of Surfaces of Revolution. 200. Examples of Surfaces of Revolution.

PART I.

THE DIFFERENTIAL CALCULUS.

USE OF THE SYMBOL ≡.

The symbol ≡ of identity as employed in this work indicates that the single letter on one side of it is used to represent the expression or thing defined on the other side of it.

When the single letter precedes the symbol ≡, the latter may commonly be read *is put for*, or *is defined as*.

When the single letter follows the symbol, the latter may be read *which let us call*.

In each case the equality of the quantities on each side of ≡ does not follow from anything that precedes, but is assumed at the moment. But having once made this assumption, any equations which may flow from it are expressed by the sign =, as usual.

PART I.

THE DIFFERENTIAL CALCULUS.

CHAPTER I.

OF VARIABLES AND FUNCTIONS.

1. In the higher mathematics we conceive ourselves to be dealing with pairs of quantities so related that the value of one depends upon that of the other. For each value which we assign to one we conceive that there is a *corresponding* value of the other.

For example, the time required to perform a journey is a function of the distance to be travelled, because, other things being equal, the time varies when the distance varies.

We study the relation between two such quantities by assigning values at pleasure to one, and ascertaining and comparing the corresponding values of the other.

The quantity to which we assign values at pleasure is called the **independent variable.**

The quantity whose values depend upon those of the independent variable is called a **function** of that variable.

EXAMPLE I. If a train travels at the rate of 30 miles an hour, and if we ask how long it will take the train to travel 15 miles, 30 miles, 60 miles, 900 miles, etc., we shall have for the corresponding times, or functions of the distances, half an hour, one hour, two hours, thirty hours, etc.

In thinking thus we consider the *distance* to be travelled as the independent variable, and the *time* as the function of the distance.

EXAMPLE II. If between the quantities x and y we have the equation
$$y = 2ax^2,$$
we may suppose
$$x = -1, 0, +1, +2, +3, \text{etc.},$$
and we shall then have
$$y = 2a, 0, 2a, 8a, 18a, \text{etc.}$$

Here x is taken as the independent variable, and y as the function of x. For each value we assign to x there is a corresponding value of y.

When the relation between the two quantities is expressed by means of an equation between symbolic expressions, the one is called an **analytic function** of the other.

An analytic function is said to be

Explicit when the symbol which represents it stands alone on one side of the equation;

Implicit when it does not so stand alone.

EXAMPLE. In the above equation y is an explicit function of x. But if we have the equation
$$y^2 + xy = x^2,$$
then for each value of x there will be a certain value of y, which will be found by solving the equation, considering y as the unknown quantity. Here y is still a function of x, because to each value of x corresponds a certain value of y; but because y does not stand alone on one side of the equation it is called an implicit function.

REMARK. The difference between explicit and implicit functions is merely one of form, arising from the different ways in which the relation may be expressed. Thus in the two forms

$$y = 2ax^2,$$
$$y - 2ax^2 = 0,$$

y is the same function of x; but its form is explicit in the first and implicit in the second.

An implicit function may be reduced to an explicit one by solving the equation, regarding the function as the unknown quantity. But as the solution may be either impracticable or too complicated for convenient use, it may be impossible to express the function otherwise than in an implicit form.

2. *Classification of Functions.* When y is an explicit function of x it is, by definition, equal to a symbolic expression containing the symbol x. Hence we may call either y or the symbolic expression the function of x, the two being equivalent. Indeed any algebraic expression containing a symbol is, by definition, a function of the quantity represented by the symbol, because its value must depend upon that of the symbol.

Every algebraic expression indicates that certain operations are to be performed upon the quantities represented by the symbols. These operations are:

1. Addition and subtraction, included algebraically in one class.

2. Multiplication, including involution.

3. Division.

4. Evolution, or the extraction of roots.

A function which involves only these four operations is called **algebraic**.

Functions are classified according to the operations which must be performed in order to obtain their values from the values of the independent variables upon which they depend.

A **rational function** is one in which the only operations indicated upon or with the independent variable are those of addition, multiplication, or division.

An **entire** function is a rational one in which the only indicated operations are those of addition and multiplication.

EXAMPLES. The expression
$$a + bx + cx^2 + dx^3$$
is an entire function of x, as well as of a, b, c and d.

The expression
$$a + \frac{m}{x} + \frac{c}{x^2 + nx}$$
is a rational function of x, but not an entire function of x.

An **irrational** function of a variable is one in which the extraction of some root of an expression containing that variable is indicated.

EXAMPLE. The expressions
$$\sqrt{a + bx}, \quad (a + mx^2 + nx^6)$$
are irrational functions of x.

Functions which cannot be represented by any finite combination of the algebraic operations above enumerated are called **transcendental**.

An **exponential** function is one in which the variable enters into an exponent.

EXAMPLE. The expressions
$$(a + x)^{ny}, \quad x^{2y}$$
are entire functions of x when n and y are integers. But they are exponential functions of y.

Other transcendental functions are:

Trigonometric functions, the sine, cosine, etc.

Logarithmic functions, which require the finding of a logarithm.

Circular functions, which are the inverse of the trigonometric functions; for example, if

$y =$ a trigonometric function of x, sin x for instance,

then x is a circular function of y, namely, the arc of which y is the sine.

3. *Functional Notation.* For brevity and generality we may represent any function of a variable by a single symbol having a mark to indicate the variable attached to it, in any form we may elect. Such a symbol is called a **functional symbol** or a **symbol of operation**.

The most common functional symbols are

$$F, \quad f \quad \text{and} \quad \phi;$$

but any signs or mode of writing whatever may be used. Then, such expressions as

$$F(x), \quad f(x), \quad \phi(x),$$

each mean

"some symbolic expression containing x."

The variable is enclosed in parentheses in order that the function may not be mistaken for the product of a quantity F, f or ϕ by x.

Identical Functions. Functions which indicate identical operations upon two variables are considered as identical.

EXAMPLE. If we consider the expression

$$a + by$$

as a certain function of y, then

$$a + bx$$

is *that same* function of x, and

$$a + b(x + y)$$

is *that same* function of $x + y$.

When the functional notation is applied, then:

Identical functions are represented by the same functional symbols.

EXAMPLES. If we put

$$F(x) \equiv a + bx,$$

we shall have
$$F(y) = a + by;$$
$$F(y^2) = a + by^2;$$
$$F(x^2 - y^2) = a + b(x^2 - y^2).$$

In general, *If we define a functional symbol as representing a certain function of a variable, that same symbol applied to a second variable will represent the expression formed by substituting the second variable for the first.*

In applying this rule any expression may be regarded as a variable to be substituted, as, in the last example, we used $x^2 - y^2$ as a variable to be substituted for x in the original expression.

EXERCISES.

1. If we put
$$\phi(x) \equiv ax^2,$$
it is required to form and reduce the functions
$$\phi(y), \quad \phi(b), \quad \phi(a), \quad \phi(-x), \quad \phi(x^2), \quad \phi\left(\frac{x}{a}\right).$$

2. Putting
$$F(x) \equiv \frac{1+x}{1-x},$$
it is required to form and reduce
$$F(x+1), \quad F\left(\frac{1}{x}\right), \quad F\left(\frac{x}{y}\right), \quad F\left(\frac{y}{x}\right), \quad F\left(\frac{a}{b}\right) + F\left(\frac{b}{a}\right).$$

3. Putting
$$f(x) \equiv \frac{x-a}{x+a},$$
it is required to form and reduce
$$f(x-a), \quad f(x+a), \quad f\left(\frac{x}{a}\right), \quad f\left(\frac{a}{x}\right).$$

4. If
$$\phi(x) \equiv a^2x + cx^2,$$
form and reduce the expressions
$$\phi(x^2), \quad \phi(a^2), \quad \phi(ax), \quad \phi(bx), \quad \phi(a+c), \quad \phi(a-c).$$

5. Suppose $\phi(x) \equiv ax^2 - a^2x$, and thence form
$$\phi(y), \qquad \phi(z), \qquad \phi(by),$$
$$\phi(x+y), \qquad \phi(x+a), \qquad \phi(x-a),$$
$$\phi(x+ay), \qquad \phi(x-ay), \qquad \phi(x^2).$$

6. Suppose $f(x) \equiv x^2$, and thence form the values of

$$f(1), \quad f(x^2), \quad f(x^3), \quad f(x^4), \quad f(x^5), \quad f(x^n).$$

7. Let us put $\phi(m) \equiv m(m-1)(m-2)(m-3)$; thence find the values of

$$\phi(6), \phi(5), \phi(4), \phi(3), \phi(2), \phi(1), \phi(0), \phi(-1), \phi(-2).$$

8. Prove that if we put $\phi(x) \equiv a^x$, we shall have

$$\phi(x+y) = \phi(x) \times \phi(y); \quad \phi(xy) = [\phi(x)]^y = [\phi(y)]^x.$$

4. *Functions of Several Variables.* An algebraic expression containing several quantities may be represented by any symbol having the letters which represent the quantities attached.

EXAMPLES. We may put

$$\phi(x, y) \equiv ax - by,$$

the comma being inserted between x and y so that their product shall not be understood. We shall then have

$$\phi(m, n) = am - bn,$$
$$\phi(y, x) = ay - bx,$$

the letters being simply interchanged;

$$\phi(x+y, x-y) = a(x+y) - b(x-y)$$
$$= (a-b)x + (a+b)y;$$
$$\phi(a, b) = a^2 - b^2;$$
$$\phi(b, a) = ab - ba = 0;$$
$$\phi(a+b, ab) = a(a+b) - ab^2;$$
$$\phi(a, a) = a^2 - ba;$$
etc. etc.

If we put $\phi(a, b, c) \equiv 2a + 3b - 5c$, we shall have

$$\phi(x, z, y) = 2x + 3z - 5y;$$
$$\phi(z, y, x) = 2z + 3y - 5x;$$
$$\phi(m, m, -m) = 2m + 3m + 5m = 10m;$$
$$\phi(3, 8, 6) = 2 \cdot 3 + 3 \cdot 8 - 5 \cdot 6 = 0.$$

EXERCISES.

Let us put
$\phi(x, y) \equiv 3x - 4y;$
$f(x, y) \equiv ax + by;$
$f(x, y, z) \equiv ax + by - abz.$

Thence form the expressions:

1. $\phi(y, x)$.
2. $\phi(a, b)$.
3. $\phi(3, 4)$.
4. $\phi(4, 3)$.
5. $\phi(10, 1)$.
6. $f(a, b)$.
7. $f(b, a)$.
8. $f(y, x)$.
9. $f(7, -3)$.
10. $f(q, -p)$.
11. $f(z, x, y)$.
12. $f(b, a, 2)$.
13. $f(a, b, c)$.
14. $f(a^2, b^2, c^2)$.
15. $f(-a, -b, -ab)$.

Sometimes there is no need of any functional symbol except the parentheses. For example, the form (m, n) may be used to indicate any function of m and n.

EXERCISES.

Let us put $(m, n) \equiv \dfrac{m(m-1)(m-2)}{n(n-1)(n-2)},$
then find the values of—

1. $(3, 3)$.
2. $(4, 3)$.
3. $(5, 3)$.
4. $(6, 3)$.
5. $(7, 3)$.
6. $(8, 3)$.
7. $(2, -1)$.
8. $(3, -2)$.
9. $(4, -2)$.

5. *Functions of Functions.* By the definitions of the preceding chapter, the expression

$$f\Big(\phi(x)\Big)$$

will mean the expression obtained by substituting $\phi(x)$ for x in $f(x)$.

We may here omit the larger parentheses and write $f\phi(x)$ instead of $f\Big(\phi(x)\Big)$.

For example, using the notation of exercises 1 and 3 of § 3, we shall have

$$f\phi(x) = \frac{ax^2 - a}{ax^2 + a} = \frac{x^2 - 1}{x^2 + 1};$$
$$\phi f(x) = a\left(\frac{x-a}{x+a}\right)^2.$$

For brevity we use the notation
$$\phi^2(x) \equiv \phi\big(\phi(x)\big).$$
Continuing the same system, we have
$$\phi^3(x) \equiv \phi\big(\phi^2(x)\big) = \phi^2\big(\phi(x)\big);$$
$$\phi^4(x) \equiv \phi\big(\phi^3(x)\big) = \phi^3\big(\phi(x)\big);$$
etc.   etc.   etc.

EXAMPLES. 1. If
$$\phi(x) \equiv ax^2,$$
then
$$\phi^2(x) = a(ax^2)^2 = a^3x^4;$$
$$\phi^3(x) = a(a^3x^4)^2 = a^7x^8;$$
etc.   etc.   etc.

2. If
$$f(x) \equiv a - x,$$
then
$$f^2(x) = a - (a - x) = x;$$
$$f^3(x) = a - f^2(x) = a - x;$$
and, in general,
$$f^{n-2}(x) = f^n(x).$$

REMARK. The functional nomenclature may be simplified to any extent.

1. The parentheses are quite unnecessary when there is no danger of mistaking the form for a product.

2. When it is once known what the variables are, we may write the functional symbol without them. Thus the symbol ϕ may be taken to mean ϕx or $\phi(x)$.

6. *Product of the First n Numbers.* The symbol $n!$, called *factorial n*, is used to express the product of the first n numbers,
$$1 \cdot 2 \cdot 3 \cdot \ldots n.$$
Thus,
$$1! = 1;$$
$$2! = 1 \cdot 2 = 2;$$
$$3! = 1 \cdot 2 \cdot 3 = 6;$$
$$4! = 1 \cdot 2 \cdot 3 \cdot 4 = 24;$$
etc.   etc.

It will be seen that
$$2! = 2 \cdot 1!;$$
$$3! = 3 \cdot 2!;$$
and, in general, $\quad n! = n \cdot (n-1)!,$
whatever number n may represent.

EXERCISES.

Compute the values of—

1. $5!$
2. $6!$
3. $8!$
4. $\dfrac{7!}{3!\,4!}$
5. $\dfrac{8!}{3!\,5!}$
6. Prove the equation $2 \cdot 4 \cdot 6 \cdot 8 \cdots 2n = 2^n n!$
7. Prove that, when n is even,
$$\frac{n}{2}! = \frac{n(n-2)(n-4)\ldots 4 \cdot 2}{2^{\frac{n}{2}}}.$$

7. *Binomial Coefficients.* The binomial coefficient
$$\frac{n(n-1)(n-2)\ldots \text{to } s \text{ terms}}{1 \cdot 2 \cdot 3 \cdots s}$$
is expressed in the abbreviated form
$$\binom{n}{s},$$
the parentheses being used to distinguish the expression from the fraction $\dfrac{n}{s}$.

EXAMPLES.

$$\binom{3}{1} = \frac{3}{1} = 3.$$
$$\binom{7}{5} = \frac{7 \cdot 6 \cdot 5 \cdot 4 \cdot 3}{1 \cdot 2 \cdot 3 \cdot 4 \cdot 5} = 21.$$
$$\binom{n}{1} = \frac{n}{1} = n.$$
$$\binom{n}{3} = \frac{n(n-1)(n-2)}{1 \cdot 2 \cdot 3}.$$

VARIABLES AND FUNCTIONS.

EXERCISES.

Prove the formulæ:

1. $\binom{5}{2} = \binom{5}{3}$.

2. $\binom{n}{2} = \left(\dfrac{n}{n-2}\right)$.

3. $\binom{5}{2} = \dfrac{5!}{2!\ 3!}$.

4. $\binom{n}{s} = \dfrac{n!}{s!\,(n-s)!}$.

5. $\binom{n+1}{s+1} = \dfrac{n+1}{s+1}\binom{n}{s}$.

6. $\binom{n}{1} + \binom{n}{2} = \left(\dfrac{n+1}{2}\right)$.

7. $\binom{n}{2} + \binom{n}{3} = \left(\dfrac{n+1}{3}\right)$.

8. $\binom{n}{3} + \binom{n}{4} = \left(\dfrac{n+1}{4}\right)$.

8. *Graphic Representation of Functions.* The methods of Analytic Geometry enable us to represent functions to the eye by means of curves. The common way of doing this is to represent the independent variable by the abscissa of a point, and the corresponding value of the function by its ordinate.

Let x_1, x_2, x_3, etc., be different values of the independent variable, and y_1, y_2, y_3, etc., the corresponding values of the function. We lay off upon the axis of abscissas the lengths OX_1, OX_2, OX_3, etc., equal

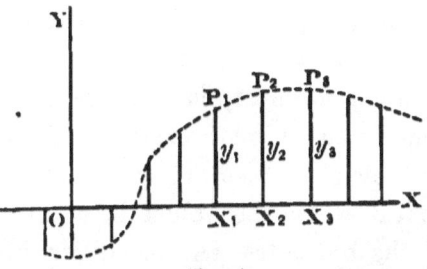

Fig. 1.

to x_1, x_2, x_3, etc., and terminating at the points X_1, X_2, X_3, etc. At each of these points we erect a perpendicular to represent the corresponding value of y. The ends, P_1, P_2, P_3, of these perpendiculars will generally terminate on a curve line, the form of which shows the nature of the function.

It must be clearly seen and remembered that it is not the curve itself which represents the values of the function, but the ordinates of the curve.

9. *Continuity and Discontinuity of Functions.* Let us consider the graphic representation of a function in the most general way. We measure off a series of values, OX_1, OX_2, OX_3, etc., of the independent variable, and at the points X_1, X_2, X_3, etc., we erect ordinates. In order that the variable ordinate may actually be a function of x it is sufficient if, for every value of the abscissa, there is a corresponding value of the ordinate.

Now we might conceive of such a function that there should be no relation between the different values of the ordinates, but that every separate point should have its own separate ordinate, as shown in Fig. 2. If this remained true how numerous soever we made the ordinates, then the ends of the latter would not terminate in any curve at all, but would be scattered over the plane. Such a function would be called *discontinuous at every point.*

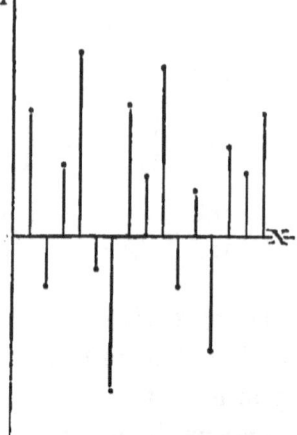

Fig. 2.

Such, however, is not the kind of functions commonly considered in mathematics. The functions with which we are now concerned are such that, however irregular they may appear when the values of x are widely separated, the ends of the ordinates will terminate in a curve when we bring those values close enough together.

If a function is such that when the point representing the independent variable moves continuously from X_1 to X_2 (Fig. 1) the end of the ordinate describes an unbroken curve, then we call the function **continuous** between the values x_1 and x_2 of the independent variable.

If the curve remains unbroken how far soever we suppose x to increase, positively or negatively, we call the function *continuous for all values of the independent variable.*

But if there is a value a of x for which there is a break of any kind in the curve, we call the function *discontinuous for the value a of the independent variable*.

Let us, for example, consider the function

$$y = \frac{a^2}{5(a-x)}.$$

Let us measure off on the axis of abscissas the length $OX = a$. Then as we make our varying ordinate approach X from the left it will increase positively without limit, and the curve will extend upwards to infinity; if we approach X from the right-hand side, the ordinate will be negative and the curve will go downwards to infinity. Thus the curve will not form a continuous branch from the one side to the other. Thus the above function is *discontinuous* for the value a of x.

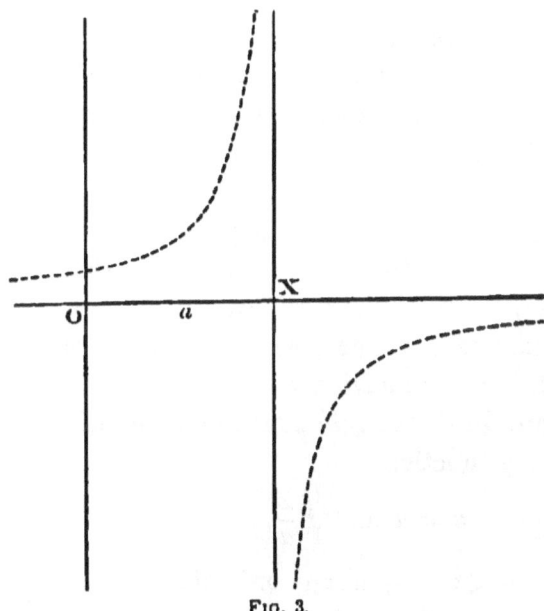

FIG. 3.

10. *Many-valued Functions.* In all that precedes, we have spoken as if to each value of the independent variable corresponded only one value of the function. But it may happen that there are several such values. For example, if y is an implicit function of x represented by the equation

$$y^3 + mxy^2 + nx^2y + px^3 = 0,$$

then we know, by the theory of equations, that there will be three values of y for each value assigned to the variable x.

Def. According as a function admits of one, two or n values, it is called one-valued, two-valued or n-valued.

Infinitely-valued Functions. It may happen that to each value of the variable there are an infinity of different values of the function. A case of this is the function $\sin^{(-1)} x$, or the arc of which x is the sine. This arc may be either the smallest arc which has x for its sine, or this smallest arc increased by any entire number of circumferences.

Take, for example, the arc whose sine shall be $+\frac{1}{2}$.

The two smallest arcs will be
$$30° = \tfrac{1}{6}\pi \quad \text{and} \quad 150° = \tfrac{5}{6}\pi.$$

But if we take the function in its most general sense it may have any of the values

$(2+\tfrac{1}{6})\pi; \quad (4+\tfrac{1}{6})\pi; \quad (6+\tfrac{1}{6})\pi, \quad \text{etc.},$

or $(2+\tfrac{5}{6})\pi; \quad (4+\tfrac{5}{6})\pi; \quad (6+\tfrac{5}{6})\pi, \quad \text{etc.}$

When we represent an n-valued function graphically, there will be n values to each ordinate. Hence each ordinate will cut the curve in n points, real or imaginary.

The figure in the margin represents the infinitely-valued function

$$y = a \sin^{(-1)} \frac{x}{a}.$$

When $-a < x < +a$, any ordinate will cut the curve in an infinity of points.

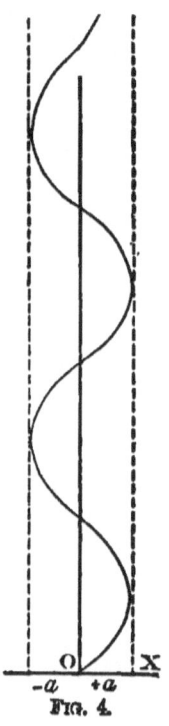

Fig. 4.

CHAPTER II.

OF LIMITS AND INFINITESIMALS.

11. *Limits.* The *method of limits* is an indirect method of arriving at the values of certain quantities which do not admit of direct determination. The method rests upon the following axioms and definition:

Axiom I. Any quantity, however small, may be multiplied by so great a number as to exceed any other quantity of the same kind, however great, to which a fixed value is assigned.

Axiom II. Conversely, any quantity, however great, may be divided into so many parts that each part shall be less than any other quantity of the same kind, however small, to which a fixed value is assigned.

Axiom III. Any quantity may be divided into any number of parts; or multiplied any number of times.

Def. The **limit** of a variable quantity X is a quantity L, which we conceive X to *approach* in such a way that the difference $L - X$ becomes less than any quantity we can name, but which we do not conceive X to *reach*.

EXAMPLE. If we have a variable quantity X and a constant quantity L, and if X, in varying according to any mathematical law, takes the successive values

$$L \pm 0.1,$$
$$L \pm 0.01,$$
$$L \pm 0.001,$$
$$L \pm 0.0001,$$

and so on indefinitely, without becoming equal to L, then we say that L is the *limit* of x.

12. *Infinites and Infinitesimals. Definitions.*

1. An **infinite** quantity is one considered as becoming greater than any quantity which we can name.

2. An **infinitesimal** quantity is one considered in the act of becoming less than any quantity which we can name; that is, in the act of approaching zero as a limit.

3. A **finite** quantity is one which is neither infinite nor infinitesimal.*

EXAMPLES. If of a quantity x we either suppose or prove

$$x > 10,$$
$$x > 100,$$
$$x > 100000,$$

and so on *without end*, then x is called an infinite quantity.

If of a quantity h we either suppose or prove

$$h < 0.1,$$
$$h < 0.001,$$
$$h < 0.00001,$$

and so on without end, then h is an infinitesimal quantity.

The preceding conceptions of limits, infinites and infinitesimals are applied in the following ways: Let us have an independent variable x, and a function of that variable which we call y.

Now, in order to apply the method of limits, we may make three suppositions respecting the value of x, namely:

1. That x approaches some finite limit.
2. That x increases without limit (i.e., is infinite).
3. That x diminishes without limit (i.e., is infinitesimal).

In each of these cases the result may be that y approaches a finite limit, or is infinite, or is infinitesimal.

---

* Strictly speaking, the words *infinite* and *infinitesimal* are both adjectives qualifying a *quantity*. But the second has lately been used also as a noun, and we shall therefore use the word *infinite* as a noun meaning infinite quantity.

For example, let us have
$$y = \frac{x+a}{x-a}.$$
Then—

When x approaches the limit a, y becomes infinite.
When x becomes infinite, y approaches the limit $+1$.
When x becomes infinitesimal, y approaches the limit -1.

The symbol $\doteq$, followed by that of zero or a finite quantity, means "approaches the limit." The symbols $\doteq \infty$ mean "increases without limit" or "becomes infinite." Hence the three last statements may be expressed symbolically, as follows:

When $x \doteq a$, then $\dfrac{x+a}{x-a} \doteq \infty$;

When $x \doteq \infty$, then $\dfrac{x+a}{x-a} \doteq +1$;

etc. etc.

The same statements are more commonly expressed thus:

$$\lim. \frac{x+a}{x-a}(x \doteq a) = \infty;$$

$$\lim. \frac{x+a}{x-a}(x \doteq \infty) = +1;$$

$$\lim. \frac{x+a}{x-a}(x \doteq 0) = -1.$$

13. *Properties of Infinite and Infinitesimal Quantities.*

THEOREM I. *The product of an infinitesimal by any finite factor, however great, is an infinitesimal.*

Proof. Let h be the infinitesimal, and n the finite factor by which it is multiplied. I say how great soever n may be, nh is also an infinitesimal. For, if nh does not become less than any quantity we can name, let α be a quantity less than which it does not become. Then if we take, as we may,

$$h < \frac{\alpha}{n}, \qquad \text{(Axiom III.)}$$

we shall have $\qquad nh < \alpha.$

That is, nh is less than α and not less than α, which is absurd.

Hence nh becomes less than any quantity we can name, and is therefore infinitesimal, by definition.

THEOREM II. *The quotient of an infinite quantity by any finite divisor, however great, is infinite.*

Proof. Let X be the infinite quantity, and n the finite divisor. If $X \div n$ does not increase beyond every limit, let K be some quantity which it cannot exceed. Then by taking

$$X > nK, \qquad \text{(Ax. III.)}$$

we shall have $\quad \dfrac{X}{n} > K;$

that is, $\dfrac{X}{n}$ greater than the quantity which it cannot exceed, which is absurd.

Hence $X \div n$ increases beyond every limit we can name when X does, and is therefore infinite when X is infinite.

THEOREM III. *The product of any finite quantity, however small, by an infinite multiplier, is infinite.*

This follows at once from Axiom I., since by increasing the multiplier we may make the product greater than any quantity we can name.

THEOREM IV. *The quotient of any finite quantity, however great, by an infinite divisor is infinitesimal.*

This follows at once from Axiom II., since by increasing the divisor the quotient may be made less than any finite quantity.

THEOREM V. *The reciprocal of an infinitesimal is an infinite, and vice versa.*

Let h be an infinitesimal. If $\dfrac{1}{h}$ is not infinite, there must be some quantity which we can name which $\dfrac{1}{h}$ does not ex-

ceed. Let K be that quantity. Because h is infinitesimal, we may have
$$h < \frac{1}{K},$$
which gives
$$\frac{1}{h} > K;$$
that is, $\frac{1}{h}$ greater than a quantity it can never exceed, which is absurd.

The converse theorem may be proved in the same way.

14. *Orders of Infinitesimals. Def.* If the ratio of one infinitesimal to another approaches a finite limit, they are called *infinitesimals of the same order.*

If the ratio is itself infinitesimal, the lesser infinitesimal is said to be *of higher order* than the other.

THEOREM VI. *If we have a series proceeding according to the powers of h,*
$$A + Bh + Ch^2 + Dh^3 + \text{etc.,}$$
in which the coefficients A, B, C, are all finite, then, if h becomes infinitesimal, each term after the first is an infinitesimal of higher order than the term preceding.

Proof. The ratio of two consecutive terms, the third and fourth for example, is
$$\frac{Dh^3}{Ch^2} = \frac{D}{C}h.$$

By hypothesis, C and D are both finite; hence $\frac{D}{C}$ is finite; hence when h approaches the limit zero, $\frac{D}{C}h$ becomes an infinitesimal (§ 13, Th. I.). Thus, by definition, the term Dh^3 is an infinitesimal of higher order than Ch^2.

Def. The orders of infinitesimals are numbered by taking some one infinitesimal as a base and calling it *an infinitesimal of the first order.* Then, an infinitesimal whose ratio to the nth power of the base approaches a finite limit is called *an infinitesimal of the nth order.*

EXAMPLE. If h be taken as the base, the term

Bh is of the first order $\therefore Bh : h$ = the finite quantity B;
Ch^2 " " second " $\therefore Ch^2 : h^2 =$ " " C;
Eh^n " " nth " $\therefore Eh^n : h^n =$ " " E.

Cor. 1. Since when $n = 0$ we have $Bh^n = Bh^0 = B$ for all values of h, it follows that an infinitesimal of the order zero is the same as a finite quantity.

Cor. 2. It may be shown in the same way that the product of any two infinitesimals of the first order is an infinitesimal of the second order.

15. *Orders of Infinites.* If the ratio of two infinite quantities approaches a finite limit, they are called infinites of the same order.

If the ratio increases without limit, the greater term of the ratio is called an infinite of *higher order* than the other.

THEOREM VII. *In a series of terms arranged according to the powers of x,*

$$A + Bx + Cx^2 + Dx^3 + \text{etc.},$$

if A, B, C, etc., are all finite, then, when x becomes infinite, each term after the first is an infinite of higher order than the term preceding.

For, the ratio of two consecutive terms is of the form $\dfrac{C}{B}x$, which becomes infinite with x (Th. III.).

Def. Orders of infinity are numbered by taking some one infinite as a base, and calling it an infinite of the first order.

Then, an infinite whose ratio to the nth power of the base approaches a finite limit is called an infinite of the nth order.

Thus, taking x as the standard, when it becomes infinite we call Bx infinite of the first order, Cx^2 of the second order, etc.

NOTE ON THE PRECEDING CHAPTERS.

In beginning the Calculus, conceptions are presented to the student which seem beyond his grasp, and methods which seem to lack rigor. Really, however, the fundamental principle of these methods is as old as Euclid, and is met with in all works on elementary geometry which treat of the area of the circle. The simplest form in which the principle appears is seen in the following case.

Let us have to compare two quantities A and B, in order to determine whether they are equal. If they are not equal, then they must differ by some quantity. If, now, taking any arbitrary quantity h, we can prove that

$$A - B < h$$

without making any supposition respecting the value of h, this will show that A and B are rigorously equal. For if they differed by the quantity α, then when h was less than α the above inequality would not hold true. But as we have been supposed to prove it for all values of h, it must be true when h is less than α. In this case h might be considered an infinitesimal, although in the Elements of Euclid it is represented on the page of the book by a figure nearly an inch square.

Infinitesimal quantities were formerly called *infinitely small*. When they were introduced by Leibnitz many able mathematicians were unable to accept them. Bishop Berkeley wrote several essays against them, in one of which he suggested that they might be called *the ghosts of departed quantities*. The following propositions are presented in the hope that they may save the student unnecessary efforts of thought in the study of this subject.

Firstly, there is no need that a quantity should be considered as *absolutely* infinite. A mathematical magnitude, considered as a quantity, must in its very nature have boundaries, because mathematics is concerned with the relation between magnitudes as greater or less, and we can compare two magnitudes as greater or less only by comparing their boundaries. An absolutely infinite magnitude, having no boundaries to compare, cannot be compared with anything.

Secondly, it is equally unnecessary to suppose the existence, either in nature or in thought, of quantities which are absolutely smaller than any finite quantity whatever.

But however small a quantity may be, there may always be another still smaller in any ratio. Hence, although it is perfectly true that no quantity can be otherwise than finite, yet it is equally true that a quantity may be less or greater than any fixed quantity we may name.

Both infinite and infinitesimal quantities are therefore essentially *indefinite*, because by considering them in the act of increasing beyond, or decreasing below, every assignable value, we do away with the very possibility of assigning values to them. They are used only as auxiliaries to lead us to a knowledge of finite quantities, and their magnitudes are never themselves the object of investigation.

The essentially indefinite nature of infinites and infinitesimals may be illustrated as follows:

If we have an equation of the form

$$x = \frac{a}{b},$$

then for every pair of *finite* values we assign to a and b there will be a definite value of x.

But if we suppose A and B to be *infinite*, and at the same time *independent of each other*, there will be no definite value to x. Considering both terms as absolutely infinite, they will have no bounds, and therefore cannot be compared in value. Considered as increasing without limit, one may be any number of times greater than the other, and thus the fraction may have any value we choose to assign it. Seeking for the value of such a fraction is like trying to answer the old question concerning the effect of an irresistible force acting upon an immovable obstacle.

CHAPTER III.

OF DIFFERENTIALS AND DERIVATIVES.

16. *Def.* An **increment** of a variable is the difference between two values of that variable.

An equivalent definition is: An *increment* is a quantity added to one value of a variable in order to obtain another value.

Notation. An increment is expressed by the symbol Δ written before the symbol of the variable.

EXAMPLE. If we have the different variables

$$x, \quad y, \quad u,$$

and the increments $\quad \Delta x, \quad \Delta y, \quad \Delta u,$

other values of the variables will be

$$x + \Delta x, \quad y + \Delta y, \quad u + \Delta u.$$

Here Δ is not a factor multiplying x, but a symbol meaning "increment of," or, in familiar language, "a little piece of."

In considering the respective increments of an independent variable, and of its function, the following five quantities come into play and are each to be clearly conceived.

1. A *value of the independent variable*, which we may take at pleasure.

2. The *corresponding value* of the function, which will be fixed by that of the independent variable.

3. ·An *increment* of the independent variable, also taken at pleasure.

4. The *corresponding increment* of the function, determined by that of the independent variable.

5. The *ratio* of these increments.

To represent these quantities, let the relation between the variable x and the function y be expressed by a curve. Let OX be one value of x, and OX' another. Let XP and $X'P'$

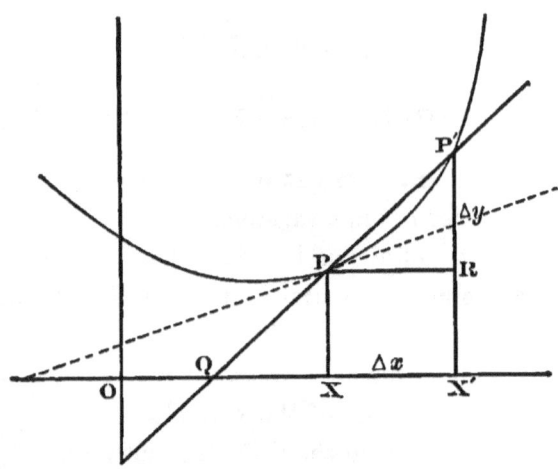

Fig. 5.

be the corresponding values of y, leading to the points P and P' of the curve. We shall then have—

1. $OX = x$, a value of the independent variable.
2. $XP = y$, the corresponding value of the function.
3. $XX' = \Delta x$, an arbitrary increment of x.
4. $RP' = \Delta y$, the corresponding increment of y.
5. Then, by Plane Trigonometry, the quotient $\dfrac{\Delta y}{\Delta x}$ will be the tangent of the angle PQX; that is, the tangent of the angle which the secant PP' makes with the axis of abscissas.

Thus we have geometrical representations of the five fundamental quantities under consideration.

17. *First Idea of Differentials and Derivatives.* Let us take, for illustration, the function

$$y = nx^2. \tag{1}$$

Giving to x the increment Δx, the new value of y will be

$$n(x + \Delta x)^2.$$

Hence $y + \Delta y = n(x + \Delta x)^2 = nx^2 + 2nx\Delta x + n\Delta x^2.$ (2)

Subtracting (1) from (2), we have, for the increment of y,
$$\Delta y = n(2x + \Delta x)\Delta x, \qquad (3)$$
Because, when Δx becomes infinitesimal,
$$\lim. (2x + \Delta x) = 2x,$$
we have, for the ratio of the increments,
$$\frac{\Delta y}{\Delta x} = 2nx + n\Delta x, \qquad (4)$$
and, when Δx becomes infinitesimal,
$$\lim. \frac{\Delta y}{\Delta x} = 2nx. \qquad (5)$$

Def. The **differential** of a quantity is its infinitesimal increment; that is, its increment considered in the act of approaching zero as its limit, or of becoming smaller than any quantity we can name.

Notation. The differential of a quantity is indicated by the symbol d written before the symbol of the quantity.

For example, the expressions
$$dx, \quad du, \quad d(x+y),$$
mean *any infinitesimal increments of* x, u, $(x+y)$, respectively.

Thus the substitution of d for Δ in the notation of increments indicates that the increment represented by Δ is supposed to be infinitesimal, and that we are to consider the limit toward which some quantity arising from the increment then approaches.

Using this notation, the equation (5) may be written
$$\frac{dy}{dx} = 2nx.$$

We also express this value of the limiting ratio in the form
$$dy = 2nx\,dx;$$
meaning thereby that the ratio of the two members of this equation has unity as its limit. This is evident from Eq. (3).

Def. If y is a function of x, the ratio $\dfrac{dy}{dx}$ of the differential of y to that of x is called the **derivative** of the function, or the **derived function**.

18. *Illustrations.* As the logic of infinitesimals offers great difficulties to the beginner, some illustrations of the subject may be of value to him.

Consider the following proposition:

The error introduced by neglecting all the powers of an increment above the first may be made as small as we please by diminishing the increment.

Let us suppose $n = 2$ in the equation (1). We then have the equations

$$\left. \begin{array}{l} y = 2x^2; \\ \Delta y = 4x\Delta x + 2\Delta x^2; \\ \dfrac{\Delta y}{\Delta x} = 4x + 2\Delta x. \end{array} \right\} \quad (a)$$

The ratio of the two terms of the second member is

$$\frac{2\Delta x}{4x}, \quad \text{or} \quad \frac{\Delta x}{2x}.$$

Let us now neglect this quantity and write the erroneous equation

$$\frac{\Delta y}{\Delta x} = 4x. \qquad (b)$$

If, now, we suppose $\left\{ \begin{array}{l} \Delta x < \dfrac{x}{100}, \\ \Delta x < \dfrac{x}{10000}, \\ \Delta x < \dfrac{x}{1000000}, \\ \text{etc.,} \end{array} \right\}$ the equation (b) will still be true within $\left\{ \begin{array}{l} \dfrac{1}{200} \text{ part;} \\ \dfrac{1}{20000} \text{ part;} \\ \dfrac{1}{2000000} \text{ part;} \\ \text{etc.} \end{array} \right.$

So long as we assign any definite value to Δx, it is clear that there will be some error in neglecting Δx. But there is no error in the equations

$$dy = 4x\,dx \quad \text{and} \quad \frac{dy}{dx} = 4x,$$

DIFFERENTIALS AND DERIVATIVES. 29

provided that we interpret them as expressing the limit which $\frac{\Delta y}{\Delta x}$ approaches as Δx approaches the limit zero, and interpret all our results accordingly.

19. *Illustration by Velocities.* Let us consider what is meant by the familiar idea of a train which may be continually changing its speed passing a certain point with a certain speed. To fix the ideas, suppose the train has just started and is every moment accelerating its speed in such manner that the total number of feet it has advanced is equal to the square of the number of seconds since it started. Put

$\delta \equiv$ the distance travelled expressed in feet;
$t \equiv$ the time expressed in seconds.

We shall then have $\delta = t^2$,

and for the distances travelled:

Number of seconds, 0; 1; 2; 3; 4; 5; etc.;
Distance travelled, 0; 1; 4; 9; 16; 25; etc.;
Distance in each second, 1; 3; 5; 7; 9; 11; etc.

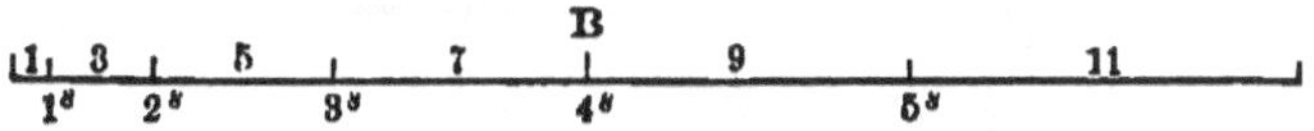

FIG. 6.

Let this line represent the space travelled the first five seconds from the starting time, and let us inquire with what velocity the train passed the point B at the end of 4^s.

Since distance travelled = velocity × time, the mean velocity is found by dividing the space by the time required to pass over that space. Now, the train had travelled

16 feet in the time 4 seconds,
and $(4 + \Delta t)^2$ feet in $(4 + \Delta t)$ seconds,
or $16 + 8\Delta t + \Delta t^2$ feet in $(4 + \Delta t)$ seconds.

Subtracting 16 feet and 4 seconds, we see that in the time Δt after the end of the 4 seconds the train went $8\Delta t + \Delta t^2$ $\equiv \Delta s$ feet. Hence its mean velocity from 4^s to $4^s + \Delta t$ is

$$\frac{\Delta s}{\Delta t} = (8 + \Delta t) \text{ feet per second.}$$

Now it is clear that, since the train was continually accelerated how small soever we take Δt, the mean velocity during this interval will exceed that with which it passed B. But it is also clear that by supposing Δt to approach the limit zero, we shall approach the required velocity as our limit. Hence the velocity with which B was passed is *rigorously*

$$\frac{ds}{dt} = 8 \text{ feet per second.}$$

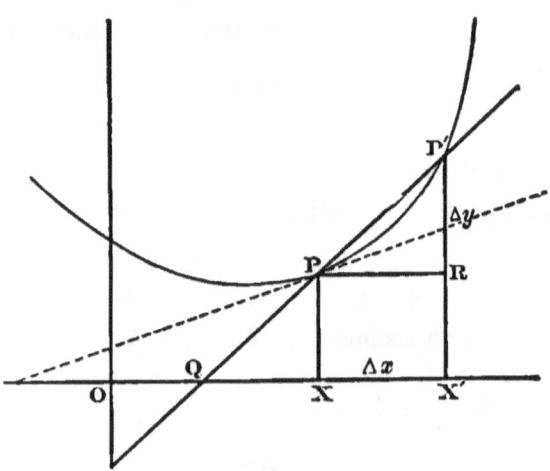

FIG. 7.

20. *Geometrical Illustration.* If, in the figure, we suppose the point P' to approach P as its limit, the increments Δx and Δy will approach the limit zero, and the secant $P'P$ will approach the tangent at the point P as its limit. We have already shown that

$$\frac{\Delta y}{\Delta x} = \text{tangent of angle made by secant with axis of abscissas.}$$

Passing to the limit, we have the rigorous proposition

$$\frac{dy}{dx} = \text{tangent of angle which the tangent at the point } P$$
makes with the axis of abscissas.

CHAPTER IV.

DIFFERENTIATION OF EXPLICIT FUNCTIONS.

21. *Def.* The process of finding the differential and the derivative of a function is called **differentiation**.

As exemplified in §§ 16, 17, it may be generalized as follows: We have given

(1) An independent variable $\equiv x$.
(2) A function of that variable $\equiv \phi(x)$.
(3) We assign to x an increment $\equiv \varDelta x$; whereby $\phi(x)$ is changed into $\phi(x + \varDelta x)$.
(4) We thus have $\phi(x + \varDelta x) - \phi(x)$ as the increment of $\phi(x)$. We may put

$$\varDelta \phi(x) \equiv \phi(x + \varDelta x) - \phi(x).$$

(5) We then form the ratio

$$\frac{\varDelta \phi(x)}{\varDelta x}, \qquad (a)$$

and seek its limit when $\varDelta x$ becomes infinitesimal. Using the notation of the last chapter, we have

$$\frac{d\phi(x)}{dx} = \lim. \frac{\varDelta \phi(x)}{\varDelta x} \; (\varDelta x \doteq 0),$$

which is the derivative of $\phi(x)$.

In order to find the ratio (a), it is necessary to develop $\phi(x + \varDelta x)$ in powers of $\varDelta x$ to at least the first power of $\varDelta x$. Let this development be

$$\phi(x + \varDelta x) = X_0 + X_1 \varDelta x + X_2 \varDelta x^2 + \ldots \qquad (1)$$

In the second member of this equation X_0, X_1, etc., will be functions of x; and it is evident that X_0 can be nothing but

$\phi(x)$ itself, because it is the value of $\phi(x + \Delta x)$ when $\Delta x = 0$. Thus we have

$$\Delta\phi(x) = \phi(x + \Delta x) - \phi(x) = (X_1 + X_2\Delta x)\Delta x + \ldots;$$
$$\frac{\Delta\phi(x)}{\Delta x} = X_1 + X_2\Delta x + \ldots.$$

Passing to the limit,

$$d\phi(x) = X_1 dx;$$
$$\frac{d\phi(x)}{dx} = X_1. \tag{2}$$

Thus, by comparing with (1), we have the following:

THEOREM I. *The derivative of a function is the coefficient of the first power of the increment of the independent variable when the function is developed in powers of that increment.*

If we have to differentiate a function of several variable quantities, x, y, z, etc., we assign an increment to each variable, and develop the function in powers and products of the increments.

Subtracting the original function, the remainder will be its increment.

The terms of highest order in this increment, considered as infinitesimals, are then called the **differential** of the function.

The following are the special cases by combining which all derivatives of rational functions may be found.

22. *Differentials of Sums.* Let x, y, z, u, etc., be any variables or functions whatever. Their sum will be

$$x + y + z + u + \text{etc.}$$

Assigning to each an increment, x will become $x + \Delta x$, y will become $y + \Delta y$, etc. Hence the sum will become

$$x + \Delta x + y + \Delta y + z + \Delta z + u + \Delta u + \text{etc.}$$

Subtracting the original expression, we find the increment of the sum to be

$$\Delta x + \Delta y + \Delta z + \Delta u + \text{etc.}$$

DIFFERENTIATION OF EXPLICIT FUNCTIONS. 33

Hence, when the increments become infinitesimal,

$$d(x + y + z + u + \text{etc.}) = dx + dy + dz + du + \text{etc.}, \quad (3)$$

or, in words:

THEOREM II. *The differential of the sum of any number of variables is equal to the sum of their differentials.*

In this theorem the quantities x, y, z, u, etc., may be either independent variables, or functions of one or more variables.

23. *Differential of a Multiple.* Let it be required to find the differential of

$$ax,$$

a being a constant.

Giving x the increment Δx, the expression will become

$$a(x + \Delta x).$$

Then, proceeding as before, we find

$$d(ax) = a\,dx. \quad (4)$$

24. THEOREM III. *The differential of any constant is zero.*

For, by definition, a constant is a quantity which we suppose invariable, and to which we cannot, therefore, assign any increment.

We therefore have, from Theorem I, when x is a variable and a is a constant,

$$d(x + a) = dx + 0 = dx,$$

or, in words:

THEOREM IV. *The differential of the sum of a constant and a variable is equal to the differential of the variable alone.*

REMARK. It will be readily seen that the conclusions of §§ 22–24 are equally true whether we suppose the increments to be finite or infinitesimal. This is no longer the case when powers or products of some finite increments enter into the expression for other finite increments.

EXERCISES.

It is required, by combining the preceding processes, to form the differentials of the following expressions, supposing a, b and c to be constants, and all the other literal symbols to be variables.

1. $u - v$.
2. $2u - v$.
3. $v + x + c$.
4. $ax + by$.
5. $a^2x + b^2y + c$.
6. $3x + 4ay + b$.
7. $4ax + 5bx - y$.
8. $6bx - abc$.
9. $3x - a + ab$.
10. $abx - abt$.
11. $c(2x + a)$.
12. $a(bx + ac)$.
13. $ac(bu + ax)$.
14. $bc(2ax - 3by)$.
15. $x - y - z$.
16. $-ax - by - cz$.
17. $-a(bx - cy)$.
18. $-b(2ax - 3cv)$.
19. $\dfrac{x}{a}$.
20. $\dfrac{x + y - z}{b}$.
21. $(a + b + c)(s + t + 3u - 4y)$.
22. $(a + 2b^2 + 3c^3)\left(\dfrac{au}{c} - \dfrac{bv}{a} + \dfrac{cw}{2} - \dfrac{abc}{3}\right)$.

25. *Differentials of Products and Powers.* Take first the product of two variables, which we shall call u and v. Then

$$\text{Product} = uv.$$

Assigning the increments Δu and Δv, the product becomes

$$(u + \Delta u)(v + \Delta v) = uv + v\Delta u + u\Delta v + \Delta u \Delta v.$$

Subtracting the original function, uv, we find

$$\Delta(uv) = v\Delta u + (u + \Delta u)\Delta v.$$

Supposing the increments to become infinitesimals, the coefficient of Δv in the second member will approach u as its limit. Hence, passing to the limit (§ 14),

$$d(uv) = vdu + udv.$$

DIFFERENTIATION OF EXPLICIT FUNCTIONS. 35

To extend the result to any number of factors, let P be the product of all the factors but one, and let the remaining factor be x, so that we have

$$\text{Product} = Px.$$

By what precedes, we have

$$d(Px) = xdP + Pdx.$$

Supposing P to be a product of the two variables u and v, this result gives

$$d(uvx) = xd(uv) + uvdx = vxdu + uxdv + uvdx. \quad (a)$$

If we add a fourth factor, y, we shall have

$$d(uvxy) = yd(uvx) + uvxdy.$$

If we substitute for $d(uvx)$ its value (a), we see that we pass from the one case to the other by (1) multiplying all the terms of the first case by the common factor y; (2) adding the product of dy into all the other factors.

We are thus led to the conclusion:

THEOREM V. *The differential of the product of any number of variables is equal to the sum of the products formed by replacing each variable by its differential.*

Corollary. If the n factors are all equal, their product will become the nth power of the variable, and the n differentials will all become equal. Hence, when n is an integer, we have the general formula

$$d(x^n) = x^{n-1}dx + x^{n-1}dx + \text{etc., to } n \text{ terms,}$$

or $\quad d(x^n) = nx^{n-1}dx.$

By combining the preceding processes we may form the differential of any entire function of any number of variables.

EXAMPLES.

1. $d(ax + bxy + cxyz)$
$= d(ax) + d(bxy) + d(cxyz) \quad\quad$ (Th. II. 22)
$= adx + bd(xy) + cd(xyz) \quad\quad$ (Th. III. 23)
$= adx + b(ydx + xdy) + c(yzdx + xzdy + xydz)$
$= (a + by + cyz)dx + (bx + cxz)dy + cxydz.$

2. $d(ax^2 + b) = d(ax^2)$ (Th. IV.)
$= ad(x^2)$ (§ 23)
$= 3ax^2 dx.$ (Th. V., Cor.)

3. $d(ax^2 y^n) = ad(x^2 y^n)$ (§ 23)
$= a[y^n d(x^2) + x^2 d(y^n)]$ (Th. V.)
$= 3ay^n x^2 dx + nax^2 y^{n-1} dy.$ (Th. V., Cor.)

4. $d(a + x^2)^n = n(a + x^2)^{n-1} d(a + x^2) = 2n(a + x^2)^{n-1} x\, dx.$

EXERCISES.

Form the differentials of the following expressions, supposing the letters of the alphabet from a to n to represent constants:

1. $a + bx^2 + cx^4.$ Ans. $(2bx + 4cx^3)dx.$
2. $B + Cy + Dy^2.$
3. $axy.$
4. $bxyz.$
5. $a(x + yz).$
6. $a(x^2 + buv).$
7. $axy + buv.$
8. $h(x^2 y + xy^2).$
9. $ax^2 y^3.$
10. $bx^2 y^n.$
11. $abx^2 y^n + ku^m v^n.$
12. $z(mx + ny).$
13. $(r + g)(t + g).$
14. $n(a - x^2).$
15. $ax^2 - byz.$
16. $(a + x)(b - y).$
17. $(a + x^2)(b - y^2).$
18. $(a - x)(a - x^2).$
19. $x(a + x)(b - x^2).$
20. $(A + Bx + Cx^2)(y + z).$
21. $(A + By^2 + Cy^4)(ay + bx).$
22. $\dfrac{xy}{a}.$
23. $(a + bu^2)(cx^2 - ny^2).$
24. $\dfrac{x - uv}{a}.$
25. $(a - x)(b - x^2)(c - x^3).$
26. $\dfrac{x - uv}{a}(u + v).$
27. $x\{x^2 + y(a - x)\}.$
28. $\left(\dfrac{x^2}{a} + \dfrac{y^2}{v}\right)xy.$
29. $(ay^2 - bx^2)(x - y).$
30. $\left(\dfrac{x}{a} - \dfrac{y}{b}\right)\left(\dfrac{x^2}{a} + \dfrac{y^2}{b}\right).$
31. $(a + x)^3.$
32. $n(a + x)^3.$
33. $(a + xy)^2.$
34. $(ax + by)^2.$

26. *Differential of a Quotient of Two Variables.* Let the variables be x and y, and let q be their quotient. Then

$$q = \frac{x}{y}$$

and $$qy = x.$$

Differentiating, we have

$$y\,dq + q\,dy = dx.$$

Solving so as to find the value of dq,

$$dq = \frac{dx - q\,dy}{y} = \frac{y\,dx - x\,dy}{y^2}.$$

Hence:

Theorem VI. *The differential of a fraction is equal to the denominator into the differential of the numerator, minus the numerator into the differential of the denominator, divided by the square of the denominator.*

Remark. If the numerator is a constant, its differential vanishes, and we have the general formula

$$d\frac{a}{x} = -\frac{a}{x^2}dx.$$

EXERCISES.

Form the differentials of the following expressions:

1. $\dfrac{x}{a+y}$.

2. $\dfrac{a+x}{a+y}$.

3. $\dfrac{a-x}{a-y}$.

4. $\dfrac{x^2}{y^2}$.

5. $\dfrac{a}{x^2}$.

6. $\dfrac{a}{(b+y)^2}$.

7. $\dfrac{a+bx}{a+by}$.

8. $\dfrac{m+nx^2}{m-nx^2}$.

9. $\dfrac{x+y}{x-y}$.

10. $\dfrac{mx^2+ny^2}{mx^2-ny^2}$.

11. $\dfrac{a}{a+bx+cx^2}$.   12. $\dfrac{x+yz}{y+xz}$.

13. $\dfrac{m+xy}{m-x^2y^2}$.   14. $\dfrac{1}{x}-\dfrac{1}{x^2}$.

15. $\dfrac{a}{x}+\dfrac{b}{y}$.   16. $\dfrac{m}{x^2}-\dfrac{n}{y^2}$.

17. $\dfrac{a}{xy+x^2y^2}$.   18. $\dfrac{\dfrac{1}{x}-\dfrac{1}{y}}{a}$.

19. $\dfrac{x^2+y^2}{x^2-y^2}$.   20. $\dfrac{x^2-y^2}{x^2+y^2}$.

27. *Differentials of Irrational Expressions.* Let it be required to find the differential of the function

$$u = x^{\frac{m}{n}},$$

m and n being positive integers. Raising both members of the equation to the nth power, we have

$$u^n = x^m.$$

Taking the differentials of both members,

$$nu^{n-1}du = mx^{m-1}dx,$$

whence

$$\frac{du}{dx} = \frac{m}{n}\frac{x^{m-1}}{u^{n-1}} = \frac{m}{n}\frac{x^{m-1}}{\left(x^{\frac{m}{n}}\right)^{n-1}} = \frac{m}{n}\frac{x^{m-1}}{x^{\frac{mn-m}{n}}} = \frac{m}{n}x^{\frac{m}{n}-1}, \; (a)$$

a formula which corresponds to the corollary of Theorem V., where the exponent is entire.

Next, let the fractional exponent be negative. Then

$$u = x^{-\frac{m}{n}} = \frac{1}{x^{\frac{m}{n}}},$$

and, by Th. VI.,

$$du = -\frac{d\left(x^{\frac{m}{n}}\right)}{x^{\frac{2m}{n}}} = -\frac{m\,x^{\frac{m}{n}-1}dx}{n\;x^{\frac{2m}{n}}} = -\frac{m}{n}x^{-\frac{m}{n}-1}dx,$$

and, for the derivative,

$$\frac{du}{dx} = -\frac{m}{n}x^{-\frac{m}{n}-1}.$$

From this equation and from (a) we conclude:

Theorem VII. *The formula*

$$d(x^n) = nx^{n-1}dx$$

holds true whether the exponent n is entire or fractional, positive or negative.

We thus derive the following rule for forming the differentials of irrational expressions:

Express the indicated roots by fractional exponents, positive or negative, and then form the differential by the preceding methods.

Examples.

1. $d\sqrt{a+x} = d(a+x)^{\frac{1}{2}} = \frac{1}{2}(a+x)^{-\frac{1}{2}}dx = \dfrac{dx}{2(a+x)^{\frac{1}{2}}}$.

2. $d\dfrac{b}{(a+x)^{\frac{1}{2}}} = d\left[b(a+x)^{-\frac{1}{2}}\right] = bd(a+x)^{-\frac{1}{2}}$

   $= -\frac{1}{2}b(a+x)^{-\frac{3}{2}}dx = -\dfrac{b}{2(a+x)^{\frac{3}{2}}}dx$.

3. $d(a+bx^2)^{\frac{1}{2}} = \frac{1}{2}(a+bx^2)^{-\frac{1}{2}}2bxdx = \dfrac{bx}{(a+bx^2)^{\frac{1}{2}}}dx$.

EXERCISES.

Form the differentials of the following expressions:

1. $\sqrt{a+x}$.   2. $\sqrt{b-x}$.   3. $\sqrt{a-bx}$.
4. $\sqrt{a-x^2}$.   5. $\sqrt{a-bx^2}$.   6. $\sqrt{x+y}$.
7. $\dfrac{a}{\sqrt{x+y}}$.   8. $\dfrac{b}{\sqrt{a+bx^2}}$.   9. $\dfrac{b}{\sqrt{a-bx^2}}$.
10. $(a+x)^{\frac{3}{2}}$.   11. $(x-a)^{\frac{3}{2}}$.   12. $(bx^2-a)^{\frac{3}{2}}$.
13. $x\sqrt{a+x}$.   14. $x\sqrt{a-x}$.   15. $y^2\sqrt{a-by^2}$.

Find the values of $\dfrac{du}{dx}$ in the following cases:

16. $u = mx + \dfrac{m}{x}$.   17. $u = (mx^2 - n)^{\frac{1}{2}}$.

18. $u = \sqrt{ax + bx^2}$. 19. $u = \dfrac{a}{b + cx^2}$.

20. $u = x\sqrt{a - x}$. 21. $u = x^2\sqrt{x^2 + a}$.

22. $u = \dfrac{a + x}{a - x}$. 23. $u = \dfrac{a - x}{a + x}$.

28. *Logarithmic Functions.* It is required to differentiate the function
$$u = \log x.$$
We have
$$\Delta u = \log(x + \Delta x) - \log x = \log \frac{x + \Delta x}{x} = \log\left(1 + \frac{\Delta x}{x}\right).$$
It is shown in Algebra that we have
$$\log(1 + h) = M(h - \tfrac{1}{2}h^2 + \tfrac{1}{3}h^3 - \text{etc.}),$$
M being the modulus of the system of logarithms employed.

Hence, puting $\dfrac{\Delta x}{x}$ for h, we find
$$\frac{\Delta u}{\Delta x} = \frac{M}{x}\left(1 - \frac{1}{2}\frac{\Delta x}{x} + \frac{1}{3}\frac{\Delta x^2}{x^2} - \text{etc.}\right),$$
and, passing to the limit,
$$du = \frac{M dx}{x}; \qquad \frac{du}{dx} = \frac{M}{x}.$$

In the Naperian system $M = 1$. In algebraic analysis, logarithms are always understood to be Naperian logarithms unless some other system is indicated. Hence we write
$$\frac{d \cdot \log x}{dx} = \frac{1}{x}; \quad d \cdot \log x = \frac{dx}{x}.$$

EXAMPLE.
$$d \cdot \log axy = \frac{d(axy)}{axy} = \frac{ax\,dy + ay\,dx}{axy} = \frac{dy}{y} + \frac{dx}{x}.$$

REMARK. We may often change the form of logarithmic

functions, so as to obtain their differentials in various ways. Thus, in the last example, we have

$$\log (axy) = \log a + \log x + \log y,$$

from which we obtain the same differential found above. The student should find the following differentials in two ways when practicable.

EXERCISES.

Differentiate:

1. $\log (a + x)$. Ans. $\dfrac{dx}{a + x}$.
2. $\log (x - a)$.
3. $\log (x^2 + b^2)$.
4. $\log (x^3 - b)$.
5. $\log mx$.
6. $\log mx^2$.
7. $\log (ax^n + b)$.
8. $\log m^x$.
9. $\log (x + y)$.
10. $\log (x - y)$.
11. $\log xy$.
12. $\log (x^2 + y^2)$.
13. $\log (a + b)^y$.
14. $\log \dfrac{x}{y}$.
15. $\log \dfrac{x + a}{y + b}$.
16. $\log \dfrac{a - x}{b - y}$.
17. $y \log x$.
18. $\log (a - x)^m$.

29. *Exponential Functions.* It is required to differentiate the function

$$u = a^x,$$

a being a constant.

Taking the logarithms of both members,

$$\log u = x \log a.$$

Differentiating, we have, by the last article,

$$d \cdot \log u = \frac{du}{u} = dx \log a.$$

Hence
$$du = u \log a\, dx = a^x \log a\, dx;$$
$$\frac{du}{dx} = a^x \log a,$$

which is the required derivative.

If a is the Naperian base, whose value is

$$e = 2.71828\ldots,$$

we have $\log a = 1$. Hence

$$\frac{d \cdot e^x}{dx} = e^x.$$

Hence the derivative of e^x possesses the remarkable property of being identical with the function itself.

EXERCISES.

Differentiate:

1. a^{2x}. Ans. $2a^{2x} \log a\, dx$.
2. a^{nx}.
3. c^{a+x}.
4. g^{a+nx}.
5. h^{mx+ny}.
6. h^{mx-y}.
7. h^{-nx}.
8. $a^x a^y$.
9. $a^x b^y$.
10. $a^{2x} b^{3y}$.
11. $a h^x b^{-2y}$.
12. e^{x+a}.
13. $e^x e^{2y}$.
14. e^{ax+by}.

30. *The Trigonometric Functions.*

The Sine. Putting h for the increment of x, we have, by Trigonometry,

$$\sin(x+h) - \sin x = 2\cos(x + \tfrac{1}{2}h)\sin \tfrac{1}{2}h.$$

Now, let h approach zero as its limit. Then,

$\sin(x+h) - \sin x$ becomes $d \sin x$;

h becomes dx, because it is the increment of x;

$\cos(x + \tfrac{1}{2}h)$ approaches the limit $\cos x$;

$\sin \tfrac{1}{2}h$ approaches the limit $\tfrac{1}{2}h$ or $\tfrac{1}{2}dx$, because when an angle approaches zero as its limit, its ratio to its sine approaches unity as its limit (Trigonometry).

Hence, passing to the limit,

$$d \cdot \sin x = \cos x\, dx.$$

The Cosine. By Trigonometry,

$$\cos(x+h) - \cos x = -2\sin(x+\tfrac{1}{2}h)\sin\tfrac{1}{2}h.$$

Hence, as in the case of the sine,

$$d\cos x = -\sin x\, dx.$$

Taking the derivatives, we have

$$\frac{d\sin x}{dx} = \cos x;$$

$$\frac{d\cdot\cos x}{dx} = -\sin x.$$

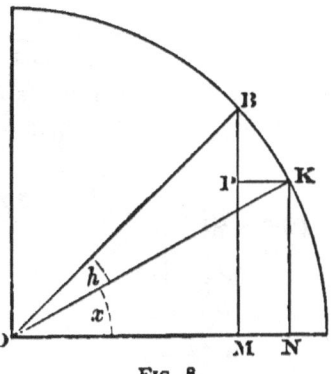

Fig. 8.

Geometrical Illustration. In the figure, let OX be the unit-radius. Then, measuring lengths in terms of this radius, we shall have

$$NK = \sin x; \quad MB = \sin(x+h); \quad \therefore\ PB = \varDelta\sin x.$$
$$ON = \cos x; \quad OM = \cos(x+h); \quad \therefore\ KP = \varDelta\cos x.$$

Also, supposing a straight line from K to B,

$$PK = -KP = KB\sin PBK;$$
$$PB = KB\cos PBK.$$

When B approaches K as its limit, the angle PBK approaches XOK, or x, as its limit, and the line KB becomes dx. Hence, approaching the limit, we find the same equations as before for $d\sin x$ and $d\cos x$.

It is evident that so long as the sine is positive, $\cos x$ diminishes as x increases, whence $d\cdot\cos x$ must have the negative sign.

The Tangent. Expressing the tangent in terms of the sine and cosine, we have

$$\tan x = \frac{\sin x}{\cos x}.$$

Differentiating this fractional expression,

$$d\tan x = \frac{\cos x\, d\cdot\sin x - \sin x\, d\cdot\cos x}{\cos^2 x} = \frac{\sin^2 x\, dx + \cos^2 x\, dx}{\cos^2 x}$$
$$= \sec^2 x\, dx.$$

We find, by a similar process,

$$d \cot x = d \cdot \frac{\cos x}{\sin x} = -\csc^2 x\, dx = -\frac{dx}{\sin^2 x};$$

$$d \cdot \sec x = d \cdot \frac{1}{\cos x} = -\frac{d \cdot \cos x}{\cos^2 x} = \frac{\sin x\, dx}{\cos^2 x}$$

$$= \tan x \sec x\, dx;$$

$$d \cdot \operatorname{cosec} x = -\cot x \csc x\, dx.$$

EXERCISES.

Differentiate:

1. $\cos (a + y)$. 2. $\sin (b - y)$. 3. $\tan (c + z)$.
4. $\sin y \cos z$. 5. $\tan u \cos v$. 6. $\sin u \tan v$.
7. $\sin ax$. 8. $\cos ay$. 9. $\tan mz$.
10. $\sin (h + my)$. 11. $\cos (h + my)$. 12. $\sin (h - my)$.
13. $\cos^2 x \cdot [d \cdot \cos^2 x = 2 \cos x\, d \cdot \cos x = -\sin 2x\, dx]$.
14. $\sin^2 x$. 15. $\sin^3 y$. 16. $\sin^3 nz$.
17. $\dfrac{\sin x}{\cos y}$. 18. $\dfrac{\sin^2 x}{\cos y}$. 19. $\dfrac{\cos^2 x}{\sin^3 y}$.

20. Show that $d(\sin^2 y + \cos^2 y) = 0$, and show why this result ought to come out by § 24.

21. Differentiate the two members of the identities

$$\cos (a + y) = \cos a \cos y - \sin a \sin y,$$
$$\sin (a + z) = \cos a \sin z + \sin a \cos z,$$

and show that the differentials of the two members of each equation are identical.

22. Show that $d \cdot \log \sin x = \cot x\, dx$;

$$d \cdot \log \cos x = -\tan x\, dx.$$

31. *Circular Functions.* A circular function is the inverse of a trigonometric function, the independent variable being the sine, cosine, or other trigonometric function, and the function the angle. The notation is as follows:

If $y = \sin z$, we write $z = \sin^{(-1)} y$ or arc-sin y;
If $u = \tan x$, we write $x = \tan^{(-1)} u$ or arc-tan u;
etc. etc. etc.

Differentiation of Circular Functions. If we have to differentiate

$$z = \sin^{(-1)} y,$$

we shall have

$$y = \sin z; \quad dy = \cos z \, dz = \sqrt{1 - \sin^2 z} \, dz;$$

$$\therefore dz = \frac{dy}{\sqrt{1 - \sin^2 z}} = \frac{dy}{\sqrt{1 - y^2}}. \qquad (a)$$

The Inverse Cosine. If z be the inverse cosine of y, we find, in the same way,

$$dz = - \frac{dy}{\sqrt{1 - y^2}}. \qquad (b)$$

The Inverse Tangent. If we have

$$z = \tan^{(-1)} y;$$

then, $y = \tan z; \quad dy = \sec^2 z \, dz = (1 + \tan^2 z) dz;$

$$\therefore dz = \frac{dy}{1 + y^2}. \qquad (c)$$

The Inverse Cotangent. We find, in a similar way,

$$d \cdot \cot^{(-1)} y = - \frac{dy}{1 + y^2}. \qquad (d)$$

The Inverse Secant. If we have

$$z = \sec^{(-1)} y;$$

then, $y = \sec z; \quad dy = \tan z \sec z \, dz = y \sqrt{y^2 - 1} \, dz;$

$$\therefore dz = \frac{dy}{y \sqrt{y^2 - 1}}. \qquad (e)$$

The Inverse Cosecant. We find, in a similar way,

$$d \cdot \csc^{(-1)} y = - \frac{dy}{y \sqrt{y^2 - 1}}.$$

EXERCISES.

Differentiate with respect to x or z:

1. $\sin^{(-1)} ax$.
2. $\cos^{(-1)}(x+a)$.
3. $\sin^{(-1)}(mx+a)$.
4. $\cos^{(-1)}\dfrac{a}{x}$.
5. $\tan^{(-1)}\left(z-\dfrac{1}{z}\right)$.
6. $\tan^{(-1)}\left(z+\dfrac{1}{z}\right)$.
7. $\tan^{(-1)}\left(\dfrac{x}{a}+\dfrac{a}{x}\right)$.
8. $\tan^{(-1)}(x^2)$.
9. $\sec^{(-1)}\left(z+\dfrac{1}{z}\right)$.
10. $\sec^{(-1)}\left(z-\dfrac{1}{z}\right)$.
11. $\sin^{(-1)} ax \cos^{(-1)}\dfrac{x}{a}$.
12. $\sec^{(-1)} x^2 \tan^{(-1)} x$.

NOTE.—The student will sometimes find it convenient to invert the function before differentiation, as we have done in deducing the differential of $\sin^{(-1)} x$.

13. We have, by comparing the above differentials,

$$d(\sin^{-1} y + \cos^{-1} y) = 0;$$
$$d(\tan^{-1} y + \cot^{-1} y) = 0;$$
$$d(\sec^{-1} y + \csc^{-1} y) = 0.$$

Show how these results follow immediately from the definition of complementary functions in trigonometry, combined with the theorem of § 24 that the differential of a constant quantity is zero.

32. *Logarithmic Differentiation.* In the case of product and exponential functions, it will often be found that the differential is most easily derived by differentiating the *logarithm* of the function. The process is then called *logarithmic differentiation.*

EXAMPLE 1. Find $\dfrac{dy}{dx}$ when $y = x^{mx}$.

We have $\qquad \log y = mx \log x;$

$$\frac{dy}{y} = m \log x \, dx + m dx;$$

$$\frac{dy}{dx} = y(m \log x + m) = m x^{mx}(1 + \log x).$$

EXAMPLE 2. $y = \dfrac{\sin^m x}{\cos^n x}$.

We have $\log y = m \log \sin x - n \log \cos x;$

$$\frac{dy}{y dx} = \frac{m \cos x}{\sin x} + \frac{n \sin x}{\cos x};$$

$$\frac{dy}{dx} = \frac{\sin^{m-1} x}{\cos^{n+1} x}(m \cos^2 x + n \sin^2 x).$$

MISCELLANEOUS EXERCISES IN DIFFERENTIATION.

Find the derivatives of the following functions with respect to x:

1. $y = x \log x.$ Ans. $\dfrac{dy}{dx} = 1 + \log x.$

2. $y = \log \tan x.$ Ans. $\dfrac{dy}{dx} = \dfrac{2}{\sin 2x}.$

3. $y = \log \cot x.$ Ans. $\dfrac{dy}{dx} = -\dfrac{2}{\sin 2x}.$

4. $y = \dfrac{x}{\sqrt{(a^2 - x^2)}}.$ Ans. $\dfrac{dy}{dx} = \dfrac{a^2}{(a^2 - x^2)^{\frac{3}{2}}}.$

5. $y = \dfrac{x^n}{(1+x)^n}.$ Ans. $\dfrac{dy}{dx} = \dfrac{n x^{n-1}}{(1+x)^{n+1}}.$

6. $y = \dfrac{e^x - e^{-x}}{e^x + e^{-x}}.$ Ans. $\dfrac{dy}{dx} = \dfrac{4}{(e^x + e^{-x})^2}.$

7. $y = \log(e^x + e^{-x}).$ Ans. $\dfrac{dy}{dx} = \dfrac{e^x - e^{-x}}{e^x + e^{-x}}.$

8. $y = \log \tan\left(\dfrac{\pi}{4} + \dfrac{x}{2}\right).$ Ans. $\dfrac{dy}{dx} = \dfrac{1}{\cos x}.$

9. $y = \dfrac{x}{e^x - 1}.$ Ans. $\dfrac{dy}{dx} = \dfrac{e^x(1-x) - 1}{(e^x - 1)^2}.$

10. $y = \dfrac{\sqrt{(1+x)} + \sqrt{(1-x)}}{\sqrt{(1+x)} - \sqrt{(1-x)}}.$ Ans. $\dfrac{dy}{dx} = -\dfrac{1}{x\sqrt{(1-x^2)}}.$

11. $y = \left\{ \dfrac{x}{1 + \sqrt{(1-x^2)}} \right\}^n$. Ans. $\dfrac{dy}{dx} = \dfrac{ny}{x\sqrt{(1-x^2)}}$.

12. $y = \tan a^{\frac{1}{x}}$. Ans. $\dfrac{dy}{dx} = -\dfrac{\sec^2 a^{\frac{1}{x}}}{x^2} \log a \cdot a^{\frac{1}{x}}$.

13. $y = x^x$. Ans. $\dfrac{dy}{dx} = x^x(1 + \log x)$.

14. $y = \sin(\log x)$. Ans. $\dfrac{dy}{dx} = \dfrac{\cos(\log x)}{x}$.

15. $y = \tan^{-1} \dfrac{x}{\sqrt{1-x^2}}$. Ans. $\dfrac{dy}{dx} = \dfrac{1}{\sqrt{1-x^2}}$.

16. $y = \log\left(\dfrac{1+x}{1-x}\right)^{\frac{1}{4}} - \dfrac{1}{2}\tan^{-1} x$.

   Ans. $\dfrac{dy}{dx} = \dfrac{x^2}{1-x^4}$.

17. $y = \log\sqrt{\dfrac{\sqrt{1+x^2}+x}{\sqrt{1+x^2}-x}}$.

   Ans. $\dfrac{dy}{dx} = \dfrac{1}{\sqrt{1+x^2}}$.

18. $y = \dfrac{1 - \tan x}{\sec x}$. Ans. $\dfrac{dy}{dx} = -(\cos x + \sin x)$.

19. $y = \log(\log x)$. Ans. $\dfrac{dy}{dx} = \dfrac{1}{x \log x}$.

20. $y = \sin^{-1} \dfrac{1 - x^2}{1 + x^2}$. Ans. $\dfrac{dy}{dx} = \dfrac{-2}{1+x^2}$.

21. $y = \log\sqrt{\dfrac{a \cos x - b \sin x}{a \cos x + b \sin x}}$.

   Ans. $\dfrac{dy}{dx} = \dfrac{-ab}{a^2 \cos^2 x - b^2 \sin^2 x}$.

22. If $y = \dfrac{1}{x}$, prove the relation $\dfrac{dy}{\sqrt{1+y^4}} + \dfrac{dx}{\sqrt{1+x^4}} = 0$.

23. $y = e^{-a^2 x^2}$. Ans. $\dfrac{dy}{dx} = -2a^2 xy$.

DIFFERENTIATION OF EXPLICIT FUNCTIONS. 49

24. $y = \dfrac{1}{(a+x)^m} \dfrac{1}{(b+x)^n}$.

    Ans. $\dfrac{dy}{dx} = - \dfrac{m(b+x) + n(a+x)}{(a+x)^{m+1}(b+x)^{n+1}}$.

25. $y = (a^2 + x^2)\tan^{-1}\dfrac{x}{a}$. Ans. $\dfrac{dy}{dx} = 2x\tan^{-1}\dfrac{x}{a} + a$.

26. $y = \sqrt{\left(\dfrac{1+x}{1-x}\right)}$. Ans. $\dfrac{dy}{dx} = \dfrac{1}{(1-x)\sqrt{1-x^2}}$.

27. $y = x + \log\cos\left(\dfrac{\pi}{4} - x\right)$. Ans. $\dfrac{dy}{dx} = \dfrac{2}{1+\tan x}$.

28. $y = x\sin^{-1}x$. Ans. $\dfrac{dy}{dx} = \sin^{-1}x + \dfrac{x}{\sqrt{1-x^2}}$.

29. $y = \tan x \tan^{-1}x$.

    Ans. $\dfrac{dy}{dx} = \sec^2 x \tan^{-1}x + \dfrac{\tan x}{1+x^2}$.

30. $y = \sin nx (\sin x)^n$.

    Ans. $\dfrac{dy}{dx} = n(\sin x)^{n-1}\sin(n+1)x$.

31. $y = \dfrac{(\sin nx)^m}{(\cos mx)^n}$.

    Ans. $\dfrac{dy}{dx} = \dfrac{mn(\sin nx)^{m-1}\cos(mx-nx)}{(\cos mx)^{n+1}}$.

32. $y = e^{-a^2x^2}\cos rx$.

    Ans. $\dfrac{dy}{dx} = -e^{-a^2x^2}(2a^2x\cos rx + r\sin rx)$.

33. $y = \log\left\{\dfrac{a + b\tan\dfrac{x}{2}}{a - b\tan\dfrac{x}{2}}\right\}$. Ans. $\dfrac{dy}{dx} = \dfrac{ab}{a^2\cos^2\dfrac{x}{2} - b^2\sin^2\dfrac{x}{2}}$.

34. $y = x^{\frac{1}{x}}$. Ans. $\dfrac{dy}{dx} = \dfrac{x^{\frac{1}{x}}(1 - \log x)}{x^2}$.

35. $y = \sin^{-1}\dfrac{x+1}{\sqrt{2}}$. Ans. $\dfrac{dy}{dx} = \dfrac{1}{\sqrt{1 - 2x - x^2}}$.

36. $y = \tan^{-1}(n\tan x)$. Ans. $\dfrac{dy}{dx} = \dfrac{n}{\cos^2 x + n^2\sin^2 x}$.

37. $y = \sec^{-1} \dfrac{a}{\sqrt{(a^2 - x^2)}}$. Ans. $\dfrac{dy}{dx} = \dfrac{1}{\sqrt{(a^2 - x^2)}}$.

38. $y = (x + a)\tan^{-1}\left(\sqrt{\dfrac{x}{a}}\right) - \sqrt{(ax)}$.

   Ans. $\dfrac{dy}{dx} = \tan^{-1}\sqrt{\dfrac{x}{a}}$.

39. $y = \sin^{-1}\sqrt{(\sin x)}$. Ans. $\dfrac{dy}{dx} = \tfrac{1}{2}\sqrt{(1 + \operatorname{cosec} x)}$.

40. $y = \tan^{-1}\dfrac{2x}{1 - x^2}$. Ans. $\dfrac{dy}{dx} = \dfrac{2}{1 + x^2}$.

41. $y = \sin^{-1}\dfrac{b + a\cos x}{a + b\cos x}$. Ans. $\dfrac{dy}{dx} = \dfrac{-\sqrt{(a^2 - b^2)}}{a + b\cos x}$.

42. $y = \cos^{-1}\dfrac{x^{2n} - 1}{x^{2n} + 1}$. Ans. $\dfrac{dy}{dx} = -\dfrac{2nx^{n-1}}{x^{2n} + 1}$.

43. $y = \sec^{-1}\dfrac{1}{2x^2 - 1}$. Ans. $\dfrac{dy}{dx} = -\dfrac{2}{\sqrt{(1 - x^2)}}$.

44. $y = \tan^{-1}\dfrac{\sqrt{(1+x^2)}-1}{x}$. Ans. $\dfrac{dy}{dx} = \dfrac{1}{2(1 + x^2)}$.

33. *Derivatives with Respect to the Time.—Velocities.* If we have a quantity which varies with the time, so as to have a definite value at each moment, but to change its value continuously from one moment to another, that quantity is, by definition, a function of the time. We now have the definition:

If we have a quantity ϕ, expressed as a function of the time $\equiv t$, the derivative, $\dfrac{d\phi}{dt}$, is the **velocity of increase**, or *rate of variation* of ϕ at any moment.

This is properly a definition of the word *velocity;* but it may be assumed that the student has already so clear a conception of what a velocity is, that he needs only to study the identity of this conception with that of a derivative relatively to t, which he can do by the illustration of § 19.

The student is recommended to draw a diagram to represent the problem whenever he can do so.

EXERCISES.

1. It is found that if a body fall in a vacuum under the influence of a constant force of gravity, the distances through which it falls in the first, second, third, fourth, etc., second of time are proportional to the numbers of the arithmetical progression

$$1, 3, 5, 7, \text{etc.},$$

or, putting a for the fall during the first second, the total fall will be

$$a + 3a + 5a + 7a + \text{etc.},$$

continued to as many terms as there are seconds. It is now required to find, by summing t terms of this progression, how far the body will fall in t seconds, and then to express its velocity in terms of t, and thus show that the velocity is proportional to the time.

Ans. (in part). The total distance fallen in t seconds will be at^2.
The velocity at the end of t seconds will be $2at$.

2. The above motion being called *uniformly accelerated*, prove this theorem: If a body fall from a state of rest with a uniformly accelerated velocity during any time τ, and if the acceleration then ceases, and the body continue with the uniform velocity then acquired, it will, during the next interval τ, fall through double the distance it did during the first interval.

Find (1) how far the body falls in τ seconds; (2) its velocity at the end of that time; (3) how far, with that velocity, it would fall in another interval of τ seconds; then show that $(3) = 2 \times (1)$.

3. The radius of a circle increases uniformly at the rate of m feet per second. At what rate per second will the area be increasing when the radius is equal to r feet?

Find (1) the expression for the value of the radius r at the end of t seconds, and (2) the area of the circle at that time. Differentiate this area, and then substitute for t its value in terms of r. Note that $\left(t = \dfrac{r}{m}\right)$.
We shall thus have $2\pi m r$ for the velocity of increase of area.

4. A body moves along the straight line whose equation is
$$x - 2y = 0$$
with a uniform velocity of n feet per second. At what rate do its abscissa and ordinate respectively increase?

Ans. $\dfrac{2n}{\sqrt{5}}$ and $\dfrac{n}{\sqrt{5}}$.

5. A man starts from a point b feet south of his door, and walks east at the rate of c feet per second. At what rate is he receding from his door at the end of t seconds?

Ans. If we put $u \equiv$ his distance from his door, we shall have
$$\frac{du}{dt} = \frac{c^2 t}{u}.$$

6. A stone is dropped from a point b feet distant in a horizontal line from the top of a flag-staff $9a$ feet high. At what rate is it receding from the top of the flag-staff (1) after it has dropped t seconds, and (2) when it reaches the ground, assuming the same law of falling as in Ex. 1?

At the end of t seconds the square of the distance from the top of the flagstaff $= u^2 = b^2 + a^2 t^4$. On reaching the ground we should have
$$\frac{du}{dt} = \frac{54 a^2}{\sqrt{b^2 + 81 a^2}}.$$

7. The sides of a rectangle grow uniformly, both starting from zero, and the one being continually double the other. Assuming one to grow at the rate of m feet and the other $2m$ feet per second, how fast will the area be growing at the end of 1, 2, 10 and t seconds? How fast, when one side is 4 and the other 8 feet?

8. The sides of an equilateral triangle increase at the rate of 2 feet per second. At what rate is the area increasing when each side is 8 feet long?

Note that the area of the triangle whose sides $= s$ is $\dfrac{\sqrt{3} s^2}{4}$.

DIFFERENTIATION OF EXPLICIT FUNCTIONS. 53

9. A man walks round a lamp, 20 feet from it, keeping the distance with a uniform motion, making one circuit per minute. Find an expression for the rate at which his shadow travels on a wall distant 40 feet from the lamp.

10. The hypothenuse of a right triangle is of the constant length of 10 feet, but slides along the sides at pleasure. If, starting from a moment when the hypothenuse is lying on the base, the end at the right angle is gradually raised up at the uniform rate of 1 foot per second, find an expression for the rate at which the other end is sliding along the base at the end of t seconds, and explain the imaginary result when $t > 10$.

11. Two men start from the same point, the one going north at the rate of 3 miles an hour, the other north-east 5 miles an hour. Find the rate at which they recede from each other.

12. A body slides down a plane inclined at an angle of 30° to the horizon, at such a rate that it has slid $3t^2$ feet at the end of t seconds. At what rates is it approaching the ground (1) at the end of t seconds, and (2) after having slid 75 feet?

13. A line revolves around the point (a, b) in the plane of a system of rectangular co-ordinate axes, making one revolution per second. Express the velocity with which its intersection with each axis moves along that axis, in terms of α, the varying angle which the line makes with the axis of X.

$$\text{Ans. } \frac{dx}{dt} = \frac{2b\pi}{\sin^2 \alpha}; \quad \frac{dy}{dt} = -\frac{2a\pi}{\cos^2 \alpha}.$$

14. A ship sailing east 6 miles an hour sights another ship 7 miles ahead sailing south 8 miles an hour. Find the rate at which the ships will be approaching or receding from each other at the end of 20, 30, 60 and 90 minutes, and at the end of t hours.

CHAPTER V.

FUNCTIONS OF SEVERAL VARIABLES AND IMPLICIT FUNCTIONS.

34. *Def.* A **partial differential** of a function of several variables is a differential formed by supposing one of the variables to change while all the others remain constant.

The **total differential** of a function is its differential when all the variables which enter into it are supposed to change.

A **partial derivative** of a function *with respect* to a quantity is its derivative formed by supposing that quantity to change while all the others remain constant.

REMARK. The adjective *partial* may be omitted when the several variables are entirely independent.

EXAMPLE. Let us have the function

$$u = x^2(y + z) + yz. \qquad (a)$$

Differentiating it with respect to x as if y and z were constant, the result will be

$$du = 2x(y + z)dx, \qquad (b)$$

which is the partial differential with respect to x. Also,

$$\left(\frac{du}{dx}\right) = 2x(y + z)$$

is the partial derivative with respect to x.

In the same way, supposing y alone to vary, we shall have

$$du = (x^2 + z)dy, \qquad (c)$$

$$\left(\frac{du}{dy}\right) = x^2 + z,$$

which are the partial differential and derivative with respect to y. For the partial differential and derivative with respect to z we have

$$du = (x^2 + y)dz; \qquad (d)$$
$$\left(\frac{du}{dz}\right) = x^2 + y.$$

Notation of Partial Derivatives. 1. A partial derivative is sometimes enclosed in parentheses, as we have done above, to distinguish it from a *total derivative* (to be hereafter defined). But in most cases no such distinctive notation is necessary.

2. In forming partial derivatives the student is recommended to use the form

$$D_x u \quad \text{instead of} \quad \frac{du}{dx},$$

because of its simplicity. It is called *the D_x of u.* The equations following (b), (c) and (d) would then be written:

$$D_x u = 2x(y + z);$$
$$D_y u = x^2 + z;$$
$$D_z u = x^2 + y.$$

EXERCISES.

Find the derivatives of the following functions with respect to x, y and z:

1. $v = x^2 - xy + y^2$.
    Ans. $D_x v = 2x - y$; $D_y v = -x + 2y$; $D_z v = 0$.
2. $w = x^3 + x^2 y + xz$.
3. $u = x^2 y^2 z^2$.
4. $u = x \log y + y \log x$.
5. $u = (x + y + z)^2$.
6. $u = \sqrt{(x + my)}$.
7. $u = (x + 2y + 3z)^{\frac{1}{2}}$.

NOTE. In forms like the last three, begin by taking the total differential, thus:

$$du = \tfrac{1}{2}(x + 2y + 3z)^{-\frac{1}{2}} d \cdot (x + 2y + 3z)$$
$$= \tfrac{1}{2}(x + 2y + 3z)^{-\frac{1}{2}} (dx + 2dy + 3dz).$$

Then, supposing x alone to vary, $D_x u = \dfrac{1}{2(x+2y+3z)^{\frac{1}{2}}}$.

supposing y alone to vary, $D_y u = \dfrac{1}{(x+2y+3z)^{\frac{1}{2}}}$.

supposing z alone to vary, $D_z u = \dfrac{3}{2(x+2y+3z)^{\frac{1}{2}}}$.

8. $w = (x + y + z)^n$.
9. $w = (x^2 + y^2 + z^2)^n$.
10. $w = \cos(mx + y)$.
11. $w = \sin(x + 2y + 3z)$.
12. $v = \tan(x - y)$.
13. $v = \sec(mx + nz)$.
14. $v = \cos^2(ax + bz)$.
15. $v = e^{x+y}$.
16. $u = xe^y + ye^x$.
17. $u = x^y + y^x$.
18. $u = \sin(x+y)\cos(x-y)$.
19. $u = x \sin y - y \sin x$.

35. Fundamental Theorem. *The total differential of a function of several variables, all of whose derivatives are continuous, is equal to the sum of its partial differentials.*

As an example of the meaning of this theorem, take the example of the preceding article, where we have found three separate differentials of u, namely, (b), (c) and (d). The theorem asserts that when x, y and z all three vary, the resulting differential of u will be the sum of these partial differentials, namely,

$$du = 2x(y + z)dx + (x^2 + z)dy + (x^2 + y)dz.$$

To show the truth of the theorem, let us first consider any function of two variables, x and y,

$$u = \phi(x, y). \qquad (1)$$

Let us now assign to x an increment $\varDelta x$, while y remains unchanged, and let us call u' the new value of u, and $\varDelta_x u$ the resulting increment of u. We shall then have

$$u' = \phi(x + \varDelta x, y); \qquad (2)$$
$$\varDelta_x u = \phi(x + \varDelta x, y) - \phi(x, y).$$

TOTAL DIFFERENTIALS. 57

In the same way, if x retains its value while y receives the increment Δy, and if we call $\Delta_y u$ the corresponding increment of u, we have

$$\Delta_y u = \phi(x, y + \Delta y) - \phi(x, y). \tag{3}$$

When Δx and Δy become infinitesimal, these increments (2) and (3) become the partial differentials with respect to x and y.

Now, to get the total increment of u, we must suppose both x and y to receive their increments. That is, instead of giving y in (1) its increment Δy, we must assign this increment in (2). Then for the increment of u we shall have, instead of (3), the result

$$\Delta_y u' = \phi(x + \Delta x, y + \Delta y) - \phi(x + \Delta x, y). \tag{4}$$

Note that (3) and (4) differ only in this: that (3) gives the value of $\Delta_y u$ *before* x has received its increment, while (4) gives $\Delta_y u$ *after* x has received its increment, and is therefore the rigorous expression for the increment of u due to Δy. Now, what the theorem asserts is that, when the increments become infinitesimal, the ratio of $\Delta_y u'$ to $\Delta_y u$ approaches unity as its limit, so that we may use (3) instead of (4). To show this, let us put

$$\phi'(x, y) \equiv \left(\frac{du}{dy}\right).$$

Then, supposing Δy to become infinitesimal, and putting $d_y u$ for that part of the differential of u arising from dy, we shall have, from (3) and (4),

$$d_y u = \phi'(x, y) dy; \tag{3'}$$
$$d_y u' = \phi'(x + \Delta x, y) dy. \tag{4'}$$

When Δx approaches zero as its limit, $\phi'(x + \Delta x, y)$ must approach the limit $\phi'(x, y)$, unless there is a discontinuity in the function ϕ', which case is excluded by hypothesis. Thus, using (3′) for (4′), we have

$$\text{Total differential of } u \equiv du = \left(\frac{du}{dx}\right)dx + \phi'(x, y)dy$$

$$= \left(\frac{du}{dx}\right)dx + \left(\frac{du}{dy}\right)dy.$$

The same reasoning may be extended to the successive cases of 3, 4, ... n variables.

The following are examples of finding some differential already considered in Chap. IV., by this more general process.

1. To differentiate $u = xy$.

$$\frac{du}{dx} = y; \quad \frac{du}{dy} = x.$$

Total differential, $\quad du = ydx + xdy.$

2. $u = \dfrac{x}{y} = xy^{-1}.$

$$\frac{du}{dx} = y^{-1}; \quad \frac{du}{dy} = -xy^{-2}dy;$$

$$du = y^{-1}dx - xy^{-2}dy = \frac{ydx - xdy}{y^2}.$$

3. $\quad u = ax + bxy + cxyz.$

$$\frac{du}{dx} = a + by + cyz;$$

$$\frac{du}{dy} = bx + cxz;$$

$$\frac{du}{dz} = cxy;$$

$$du = (a + by + cyz)dx + (bx + cxz)dy + cxydz,$$

as in § 25, Example 1.

EXERCISES.

Write the total differentials of the functions given in the exercises of § 34.

36. *Principles Involved in Partial Differentiation.* All the processes of the present chapter are aimed at the following object: Any derivative expression, such as

$$\frac{du}{dx}, \text{ or } D_x u,$$

presupposes (1) that we have the quantity u given, really or ideally, as an explicit function of x, and perhaps of other quantities; (2) that we are to get the result of differentiating this function according to the rules of Chap. IV., supposing all the quantities except x to be constant.

Now, because it is often difficult or impossible to find u as an explicit function of x, we want rules for finding the values of $D_x u$, which we could get if we had u given as such a function of x. For example, we might be able to find the equation $u = \phi(x)$ if we could only solve one or more algebraic equations. If, for any reason, we will not or cannot solve these equations, we may still find $D_x u$ whenever the equations would suffice to give u as a function of x if we only did solve them. The following articles show how this is done in all usual cases.

37. *Differentiation of Implicit Functions.* Let the relation between y and x be given by an equation of the form

$$\phi(x, y) = 0. \tag{a}$$

Representing this function of x and y by ϕ, simply, and supposing for the moment that x and y are independent variables, so that ϕ need not be zero, we shall have, by the last section,

$$d\phi = \frac{d\phi}{dx}dx + \frac{d\phi}{dy}dy.$$

But, introducing the condition that equation (*a*) must be satisfied, $d\phi$ must be zero, because x and y must so vary as to keep ϕ constantly zero. We then find, from the last equation,

$$\frac{dy}{dx} = -\frac{\dfrac{d\phi}{dx}}{\dfrac{d\phi}{dy}} = -\frac{D_x\phi}{D_y\phi}, \qquad (1)$$

which is the required form in the case of an implicit function of one variable.

Cor. If from an equation of the form $x = f(y)$ we want to derive the value of D_xy, we have

$$\phi(x, y) = x - f(y) = 0;$$
$$\frac{d\phi}{dx} = 1; \quad \frac{d\phi}{dy} = -\frac{df(y)}{dy} = -\frac{dx}{dy}.$$

Hence $\quad \dfrac{dy}{dx} = \dfrac{1}{\dfrac{dx}{dy}}.$

EXAMPLE. To find D_xy from the equation

$$\phi(x, y) = y - ax = 0.$$

We have $\quad \dfrac{d\phi}{dx} = -a; \quad \dfrac{d\phi}{dy} = 1; \quad \dfrac{dy}{dx} = a;$

the same result which we should get by differentiating the equivalent equation $y = ax$.

REMARK. If we should reduce the middle member of (1) by clearing of fractions, the result would be the negative of the correct one. This illustrates the fact that there is no relation of equality between the two differentials of each of the quantities x, y and ϕ, all that we are concerned with being the *limiting ratios* $dy : dx$; $d\phi : dx$, and $d\phi : dy$, which limiting ratios are functions of x and y.

We may, indeed, if we choose, suppose the two dx's equal and the two dy's equal. But in this case the two $d\phi$'s must have opposite algebraic signs, because their sum, or the total differential of ϕ, is necessarily zero. Now, if we change the sign of either of the $d\phi$'s, we shall get a correct result by a fractional reduction.

EXERCISES.

Find the values of $\dfrac{dy}{dx}$, $\dfrac{dx}{dz}$ or $\dfrac{du}{dx}$ from the following equations:

1. $y^2 - ax = 0$.
2. $y^2 - yx + x^2 = 0$.
3. $x^2 + 4xz + z^2 = 0$.
4. $u(a-x) + u^2(b+x) = 0$.
5. $\log x + \log y = c$.
6. $\log (x+y) + \log (x-y) = c$.
7. $\sin x + \sin y = c$.
8. $\sin ax - \sin by = c$.
9. $u + e \sin u = x$.
10. $x(1 - e \cos z) = a$.

38. *Implicit Functions of Several Variables.* The preceding process may be extended to the case of an implicit function of any number of variables in a way which the following example will make clear.

Let u be expressed as a function of x, y and z by the equation

$$u^3 + xu^2 + (x^2 + y^2)u + x^3 + y^3 + z^3 = 0.$$

Since this expression is constantly zero, its total differential is zero. Forming this total differential, we have

$$(3u^2 + 2xu + x^2 + y^2)du + (u^2 + 2ux + 3x^2)dx \\ + (2uy + 3y^2)dy + 3z^2 dz = 0.$$

By § 34 we obtain the derivative of u with respect to x by supposing all the other variables constant; that is, by putting $dy = 0$, $dz = 0$, and so with y and z. Hence

$$\frac{du}{dx} = D_x u = -\frac{u^2 + 2ux + 3x^2}{3u^2 + 2ux + x^2 + y^2};$$

$$\frac{du}{dy} = D_y u = -\frac{2uy + 3y^2}{3u^2 + 2ux + x^2 + y^2};$$

$$\frac{du}{dz} = D_z u = -\frac{3z^2}{3u^2 + 2ux + x^2 + y^2}.$$

EXERCISES.

Find the derivatives of u, v or r with respect to x, y and z from the following equations:

1. $xu^3 + y^2u^2 + z^3u = x^3yz$.
2. $a \cos(x - u) + b \sin(x + u) = y$.
3. $u^x + u^y = u^z$.
4. $r^{x+y} + r^{x-y} = r^z$.
5. $v \log x + z \log v = y$.
6. $c^r \cos x + c^x \cos y = c^y$.
7. $u^2 - 2ux \cos z + x^2 = a^2$.
8. $v^2 + 2vx \cos z + x^2 = b^2$.

39. *Case of Implicit Functions expressed by Simultaneous Equations.* If we have two equations between more than two variables, such as

$$F_1(u, v, x, y, \text{etc.}) = 0, \quad F_2(u, v, x, y, \text{etc.}) = 0,$$

then, if values of all but two of these variables are given, we may, by algebraic methods, determine the values of the two which remain. We may therefore regard these two as functions of the others, the partial derivatives of which admit of being found.

In general, suppose that we have n independent variables, $x_1, x_2 \ldots x_n$, and m other quantities, $u_1, u_2 \ldots u_m$, connected with the former by m equations of the form

$$\left. \begin{array}{l} F_1(u_1, u_2 \ldots u_m, x_1, x_2 \ldots x_n) = 0; \\ F_2(u_1, u_2 \ldots u_m, x_1, x_2 \ldots x_n) = 0; \\ \quad \cdot \quad \cdot \quad \cdot \quad \cdot \quad \cdot \quad \cdot \\ \quad \cdot \quad \cdot \quad \cdot \quad \cdot \quad \cdot \quad \cdot \\ \quad \cdot \quad \cdot \quad \cdot \quad \cdot \quad \cdot \quad \cdot \\ F_m(u_1, u_2 \ldots u_m, x_1, x_2 \ldots x_n) = 0. \end{array} \right\} \quad (a)$$

By solving these m equations (were we able to do so) we should obtain the m u's in terms of the n x's in the form

$$\left. \begin{array}{l} u_1 = \phi_1(x_1, x_2 \ldots x_n); \\ \quad \cdot \quad \cdot \quad \cdot \quad \cdot \\ \quad \cdot \quad \cdot \quad \cdot \quad \cdot \end{array} \right\} \quad (b)$$

DIFFERENTIATION OF IMPLICIT FUNCTIONS. 63

and by differentiating these equations (*b*) we should find the mn values of the derivatives $\dfrac{du_1}{dx_1}$; $\dfrac{du_1}{dx_2}$; ... $\dfrac{du_2}{dx_1}$; etc.

Now, the problem is to find these same derivatives from (*a*) without solving (*a*).

The method of doing this is to form the complete differential of each of the given equations (*a*), and then to solve the equations thus obtained with respect to du_1, du_2, etc.

The results of the differentiation may, by transposition, be written in the form

$$\dfrac{dF_1}{du_1} du_1 + \dfrac{dF_1}{du_2} du_2 + \ldots + \dfrac{dF_1}{du_m} du_m = -\dfrac{dF_1}{dx_1} dx_1 - \text{etc.};$$

$$\dfrac{dF_2}{du_1} du_1 + \dfrac{dF_2}{du_2} du_2 + \ldots + \dfrac{dF_2}{du_m} du_m = -\dfrac{F_d}{dx_1} dx_1 - \text{etc.};$$

. . . .
. . . .
. . . .

$$\dfrac{dF_m}{du_1} du_1 + \dfrac{dF_m}{du_2} du_2 + \ldots + \dfrac{dF_m}{du_m} du_m = -\dfrac{dF_m}{dx_1} dx_1 - \text{etc.}$$

By solving these m equations for the m unknown quantities du_1, du_2 ... du_m, we shall have results of the form

$$du_1 = M_1 dx_1 + M_2 dx_2 + \ldots + M_n dx_n;$$
$$du_2 = N_1 dx_1 + N_2 dx_2 + \ldots + N_n dx_n;$$
etc.   etc.   etc.   etc.;

where M_1, N_1, etc., represent the functions of $u_1, \ldots u_m$, $x_1, \ldots x_n$, which are formed in solving the equations.

We then have for the partial derivatives

$$\dfrac{du_1}{dx_1} = M_1; \quad \dfrac{du_1}{dx_2} = M_2; \quad \text{etc.}$$

EXAMPLE. From the equations

$$\left. \begin{array}{l} r \cos \theta = x, \\ r \sin \theta = y, \end{array} \right\} \qquad (a')$$

it is required to find the derivatives of r and θ with respect to x and y.

By differentiation we obtain

$$\cos\theta dr - r\sin\theta d\theta = dx;$$
$$\sin\theta dr + r\cos\theta d\theta = dy.$$

Multiplying the first equation by $\cos\theta$ and the second by $\sin\theta$, and adding, we eliminate $d\theta$. Multiplying the first by $-\sin\theta$ and the second by $\cos\theta$, and adding, we eliminate dr. The resulting equations are

$$dr = \cos\theta dx + \sin\theta dy;$$
$$rd\theta = \cos\theta dy - \sin\theta dx.$$

Hence, as in the last section,

$$\left(\frac{dr}{dx}\right) = \cos\theta; \qquad \left(\frac{dr}{dy}\right) = \sin\theta;$$
$$\left(\frac{d\theta}{dx}\right) = -\frac{\sin\theta}{r}; \qquad \left(\frac{d\theta}{dy}\right) = \frac{\cos\theta}{r}.$$

EXERCISES.

1. From the equations

$$r\sin\theta = x - y,$$
$$r\cos\theta = x + y,$$

find the derivatives of r and θ with respect to x and y.

2. From the equations

$$ue^v = r\cos\theta,$$
$$ue^{-v} = r\sin\theta,$$

find the derivatives of u and v with respect to r and θ.

Ans. $\left(\dfrac{du}{dr}\right) = \tfrac{1}{2}(e^v\sin\theta + e^{-v}\cos\theta);$

$\left(\dfrac{du}{d\theta}\right) = \dfrac{r}{2}(e^v\cos\theta - e^{-v}\sin\theta);$

$\left(\dfrac{dv}{dr}\right) = \dfrac{1}{2u}(e^{-v}\cos\theta - e^v\sin\theta);$

$\left(\dfrac{dv}{d\theta}\right) = \dfrac{r}{2u}(e^{-v}\sin\theta + e^v\cos\theta).$

3. From the equations

$$u^2 + ru = x^2 + y^2,$$
$$u^2 - ru = xy,$$

find the derivatives of r and u with respect to x and y.

4. From the equations

$$x^2 + y^2 + z^2 - 2xyz = 0,$$
$$x + y + z = a,$$

find $\dfrac{dz}{dx}$ and $\dfrac{dz}{dy}$.

5. From

$$u^2 - 2uz \cos\theta + z^2 = a^2,$$
$$w^2 + 2uz \cos\theta + z^2 = b^2,$$

find $\dfrac{du}{dz}$; $\dfrac{du}{d\theta}$; $\dfrac{dw}{dz}$; $\dfrac{dw}{d\theta}$.

40. *Functions of Functions.* Let us have an equation of the form

$$u = f(\phi, \psi, \theta, \text{etc.}); \qquad (a)$$

where ϕ, ψ, θ, etc., are all functions of x, admitting of being expressed in the form

$$\phi = f_1(x); \quad \psi = f_2(x); \quad \theta = f_3(x); \quad \text{etc.} \qquad (b)$$

If any definite value be assigned to x, the values of ϕ, ψ, θ, etc., will be determined by (b). By substituting these values in (a), u will also be determined. Hence the equations (a) and (b) determine u as a function of x.

By substituting in (a) for ϕ, ψ, θ, etc., their algebraic expressions $f_1(x)$, $f_2(x)$, etc., we shall have u as an explicit function of x, and can hence find its derivative with respect to x. But what we want to do is to find an expression for this derivative without making this substitution.

By differentiating (a) we have

$$du = \frac{du}{d\phi}d\phi + \frac{du}{d\psi}d\psi + \frac{du}{d\theta}d\theta + \text{etc.}$$

By differentiating (b),

$$d\phi = \frac{d\phi}{dx}dx; \quad d\psi = \frac{d\psi}{dx}dx; \quad d\theta = \frac{d\theta}{dx}dx; \quad \text{etc.}$$

By substituting these values in the last equation and dividing by dx, we have

$$\frac{du}{dx} = \frac{du}{d\phi}\frac{d\phi}{dx} + \frac{du}{d\psi}\frac{d\psi}{dx} + \frac{du}{d\theta}\frac{d\theta}{dx} + \text{etc.} \quad (1)$$

The significance of this equation is this: a change in x changes u in as many ways as there are functions ϕ, ψ, θ, etc.

$\dfrac{du}{d\phi}\dfrac{d\phi}{dx}dx$ is the change in u through ϕ;

$\dfrac{du}{d\psi}\dfrac{d\psi}{dx}dx$ is the change in u through ψ;

etc.   etc.

The total differential is the sum of all these separate infinitesimal changes, and the derivative is the quotient of this total differential by dx.

EXERCISES.

1. Find $\dfrac{du}{dx}$ from the equations

$$u = a \sin(mv + w) + b \sin(mv - w);$$
$$v = c + nx; \quad w = c - nx.$$

We find $\dfrac{du}{dv} = am \cos(mv + w) + bm \cos(mv - w); \quad \dfrac{dv}{dx} = n;$

$\dfrac{du}{dw} = a \cos(mv + w) - b \cos(mv - w); \quad \dfrac{dw}{dx} = -n;$

whence, by the general formula,

$$\frac{du}{dx} = an(m-1)\cos(mv+w) + bn(m+1)\cos(mv-w).$$

2. Find $\dfrac{du}{dx}$ from

$$u = e^\phi + e^\psi;$$
$$\phi = e^x; \quad \psi = ne^{-x}. \quad \text{Ans. } e^{\phi+x} - ne^{\psi-x}.$$

3. Find $\dfrac{dv}{dy}$ from

$$v^2 + v\phi + \psi^2 = a;$$
$$\phi = m(a+y); \quad \psi = ny.$$

4. Find $\dfrac{dr}{dz}$ from

$$r \cos x - r \sin x = a - y;$$
$$x = mz + h; \quad y = \cos nz.$$

5. Find $\dfrac{dr}{dz}$ from

$$r^3 + xr^2 + y^2r + \phi^3 = 0;$$
$$x^2 + az = 0; \quad y^2 + az^2 = 0; \quad \phi = nz.$$

41. The foregoing theory applies equally to the case in which the function is one of two or more variables, some of which are functions of the others. For example, if

$$u = \phi(x, z), \qquad (a)$$

then, whatever be the relation between x and z, we shall always have, for the complete differential of u,

$$du = \left(\dfrac{du}{dx}\right)dx + \left(\dfrac{du}{dz}\right)dz.$$

Suppose that x is itself a function of z. We then have

$$dx = \dfrac{dx}{dz}dz.$$

By substitution in the first equation we have

$$du = \left[\left(\dfrac{du}{dx}\right)\dfrac{dx}{dz} + \left(\dfrac{du}{dz}\right)\right]dz;$$
$$\therefore \dfrac{du}{dz} = \left(\dfrac{du}{dx}\right)\dfrac{dx}{dz} + \left(\dfrac{du}{dz}\right). \qquad (b)$$

The two values of $\dfrac{du}{dz}$ which enter into this equation are different quantities. A change in z produces a change in u in two ways: first, *directly*, through the change in z as it appears in (a); second, *indirectly*, by changing the values of x in (a). The first change depends upon $\left(\dfrac{du}{dz}\right)$ in the second member of (b); the second upon $\left(\dfrac{du}{dx}\right)\dfrac{dx}{dz}$; while the first member of (b) expresses the total change.

It is in distinguishing the two values of a derivative thus obtained that the terms *partial derivative* and *total derivative* become necessary. If we have a function of the form

$$u = f(x, y, w \ldots z),$$

in which any or all of the quantities x, y, w, etc., may be functions of z, then the *partial* derivative of u with respect to z means the derivative when we take no account of the variations of x, y, w, etc.; and the total derivative, with respect to z, is the derivative when all these variations are taken into account.

In such cases the partial derivative has to be distinguished by being enclosed in parentheses (§ 34). This is why the last equation is written

$$\frac{du}{dz} = \left(\frac{du}{dz}\right) + \left(\frac{du}{dx}\right)\frac{dx}{dz}.$$

42. *Extension of the Principle.* The principle involved in the preceding discussion may be extended to the case of any number of independent variables and any number of functions. If we have

$$r = \phi(u, v, w \ldots x, y, z \ldots),$$

in which x, y, z, etc., are the independent variables, while u, v, w, etc., are functions of these variables, we shall have

$$dr = \left(\frac{d\phi}{du}\right)du + \left(\frac{d\phi}{dv}\right)dv + \ldots + \left(\frac{d\phi}{dx}\right)dx + \text{etc.}$$

Then, since u, v, w, etc., are functions of x, y, z, etc., we have

$$du = \frac{du}{dx}dx + \frac{du}{dy}dy + \text{etc.};$$

$$dv = \frac{dv}{dx}dx + \frac{dv}{dy}dy + \text{etc.}$$

FUNCTIONS OF FUNCTIONS.

By substituting these values in the preceding equation we find*

$$dr = \left[\left(\frac{d\phi}{dx}\right) + \left(\frac{d\phi}{du}\right)\frac{du}{dx} + \left(\frac{d\phi}{dv}\right)\frac{dv}{dx} + \ldots\right]dx$$
$$+ \left[\left(\frac{d\phi}{dy}\right) + \left(\frac{d\phi}{du}\right)\frac{du}{dy} + \left(\frac{d\phi}{dv}\right)\frac{dv}{dy} + \ldots\right]dy$$
$$+ \cdot \quad \cdot \quad \cdot \quad \cdot \quad \cdot \quad \cdot$$

Hence, writing r for ϕ, its equivalent,

$$\frac{dr}{dx} = \left(\frac{dr}{dx}\right) + \left(\frac{dr}{du}\right)\frac{du}{dx} + \left(\frac{dr}{dv}\right)\frac{dv}{dx} + \text{etc.};$$

etc.   etc.   etc.   etc.

EXERCISES.

The independent variables r and θ being connected with x and y by the equations

$$x = r \cos \theta,$$
$$y = r \sin \theta,$$

it is required to find the derivatives of the following functions of x, y, r and θ with respect to r and θ. We call each of the functions u.

1. $u = r^2 + 2xy \cos 2\theta.$

Here we have

$$\left(\frac{du}{dr}\right) = 2r; \qquad \left(\frac{du}{d\theta}\right) = -4xy \sin 2\theta;$$

$$\frac{du}{dx} = 2y \cos 2\theta; \qquad \frac{du}{dy} = 2x \cos 2\theta;$$

$$\frac{dx}{dr} = \cos \theta; \qquad \frac{dy}{dr} = \sin \theta;$$

$$\frac{dx}{d\theta} = -r \sin \theta = -y; \qquad \frac{dy}{d\theta} = r \cos \theta = x.$$

---

* Here, when we use the symbol ϕ instead of r, there is really no need of enclosing the partial derivatives in parentheses. We have done it only for the convenience of the student.

Hence $\dfrac{du}{dr} = \left(\dfrac{du}{dr}\right) + \dfrac{du}{dx}\dfrac{dx}{dr} + \dfrac{du}{dy}\dfrac{dy}{dr}$

$= 2r + 2y \cos\theta \cos 2\theta + 2x \sin\theta \cos 2\theta$
$= 2r(1 + \cos 2\theta \sin 2\theta) = r(2 + \sin 4\theta);$

and, in the same way,

$$\dfrac{du}{d\theta} = 2r^2 \cos 4\theta.$$

We might have got the same result, and that more simply, by substituting for x and y in the given equation their values in terms of r and θ. But in the case of implicit functions this substitution cannot be made; it is therefore necessary to be familiar with the above method.

2. $u = \dfrac{a^2}{r^2} + \dfrac{x^2 - y^2}{a^2} \cos 2\theta.$

3. $u = \dfrac{a^2}{x^2} + \dfrac{b^2}{y^2} - \dfrac{2ab}{r^2}.$

4. $u = r^2 - (x - y)^2.$

5. $u = \dfrac{1}{x \sin 2\theta + y \cos 2\theta}.$

6. $u = \dfrac{1}{x \cos 2\theta} - \dfrac{1}{y \sin 2\theta}.$

7. $u = r^3 + x^3 - y^3.$

Let v and w be given as implicit functions of ρ and θ by the equations

$$\left.\begin{array}{r}w = av; \\ v^2 + w^2 = 2\rho \sin\theta.\end{array}\right\} \quad \cdots \cdots (a)$$

It is required to find the total derivatives of the following functions with respect to ρ and θ respectively:

8. $u = v^2 + w^2 - \rho^2.$ 9. $u = v^2 - 2vw \cos\theta + w^2.$

10. $u = \dfrac{ab}{vw}.$ 11. $u = (v + w) \sin\theta.$

12. $u = (v - w) \cos\theta.$

13. $u = w^2 - v^2 + 2(w + v)\rho \cos\theta.$

From the pair of equations (*a*) we find

$$\frac{dv}{d\rho} = \frac{v}{2\rho}; \quad \frac{dw}{d\rho} = \frac{w}{2\rho};$$

$$\frac{dv}{d\theta} = \tfrac{1}{2}v \cot \theta; \quad \frac{dw}{d\theta} = \tfrac{1}{2}w \cot \theta;$$

which values are to be substituted in the symbolic partial derivatives of *u*.

43. *Remarks on the Nomenclature of Partial Derivatives.* There is much diversity among mathematicians in the nomenclature pertaining to this subject. Thus, the term "partial derivative" is sometimes extended to all cases of a derivative of a function of several variables, with respect to any one of those variables, though there is then nothing to distinguish it from a total derivative.

Again, Jacobi and other German writers put the total derivatives in parentheses and omit the latter from the partial ones, thus reversing the above notation.

If we have to express the derivative of $\phi(x, y, z,$ etc.) with respect to z, the English writers commonly use the symbol $\dfrac{d}{dz}$ in order to avoid writing a cumbrous fraction. We thus have such forms as

$$\frac{d}{dx}\left(\frac{x^2}{a^2} + \frac{xy}{b^2} + \frac{y^2}{c^2}\right);$$

$$D_x\left(\frac{x^2}{a^2} + \frac{xy}{b^2} + \frac{y^2}{c^2}\right);$$

each of which means the derivative of the expression in parentheses with respect to x, and which the student can use at pleasure.

44. *Dependence of the Derivative upon the Form of the Function.* Let x and y be two variables entirely independent of each other, and

$$u = \phi(x, y) \qquad (a)$$

a function of these variables. Without making any change in u or x, let us introduce, instead of y, another independent variable, z, supposed to be a function of x and y. Then, after making the substitution, we shall have a result of the form

$$u = F(x, z). \qquad (b)$$

Now, it is to be noted that although both u and x have the same meaning in (b) as in (a), the value of $\dfrac{du}{dx}$ will be different in the two cases. The reason is that in (a) y is supposed constant when we differentiate with respect to x, while in (b) it is z which is supposed constant.

Analytic Illustration. Let us have

$$u = ax^2 + by^2.$$

This gives
$$\dfrac{du}{dx} = 2ax. \qquad (c)$$

Let us now substitute for y another quantity, z, determined by the equation

$$z = y + x \quad \text{or} \quad y = z - x.$$

We then have
$$u = ax^2 + b(z - x)^2;$$
$$\dfrac{du}{dx} = 2ax + 2b(x - z);$$

which is different from (c).

Our general conclusion is: *The partial derivative of one variable with respect to another depends not only upon the relation of those two variables, but upon their relations to the variables which we suppose constant in differentiating.*

Geometrical Illustration. Let r and θ be the polar co-ordinates of a point P, and x and y its rectangular co-ordinates. Then

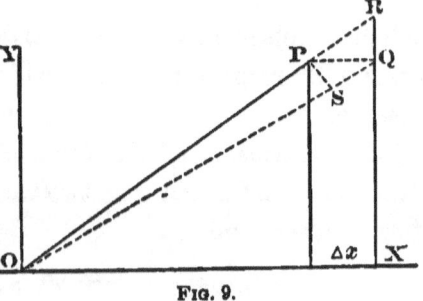

Fig. 9.

$$x = r \cos \theta;$$
$$y = r \sin \theta;$$
$$r^2 = x^2 + y^2. \qquad (d)$$

PARTIAL DERIVATIVES.

Regarding r as a function of x and y, we have

$$\frac{dr}{dx} = \frac{x}{r} = \cos \theta. \qquad (e)$$

But we may equally express r as a function of x and θ, thus:

$$r = x \sec \theta. \qquad (f)$$

We then have $\quad \dfrac{dr}{dx} = \sec \theta. \qquad (g)$

Referring to the figure, it will be seen that we derive (e) from (d) by supposing x to vary while y remains constant; that is, by giving the point P an infinitesimal motion along the line $PQ \parallel$ to OX. In this case it is plain that the increment of r (SQ) is less than that of x. But in deriving (g) from (f) we suppose x to vary while θ remains constant. This carries the point P along the straight line OPR; and now it is evident that the resulting increment of r (PR) is greater than that of x.

CHAPTER VI.

DERIVATIVES OF HIGHER ORDERS.

45. If we have given a function of x,

$$y = \phi(x),$$

we may, by differentiation, find a value of $\dfrac{dy}{dx}$. This value will, in general, be another function of x, which we may call $\phi'(x)$. Thus we shall have

$$\frac{dy}{dx} = \phi'(x).$$

Now, this function ϕ' may itself be differentiated. If we call its derived function ϕ'', we shall have

$$\frac{d\frac{dy}{dx}}{dx} = \frac{d\phi'(x)}{dx} = \phi''(x). \quad (a)$$

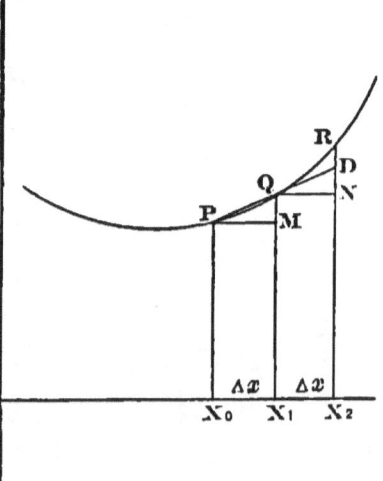

Fig. 10.

Let us examine the geometrical meaning of this equation, by plotting the curve representing the original equation $y = \phi(x)$.

Let x, x' and x'' be three equidistant values of the abscissa, so that the increments $x' - x$ and $x'' - x' \equiv \varDelta x$ are equal. Let P, Q and R be the corresponding points of the curve. Let y, y' and y'' be the three corresponding values of y.

Then we may put
$$\Delta y \equiv y' - y = MQ,$$
$$\Delta' y \equiv y'' - y' = NR,$$
as the two corresponding increments of y.

It is evident that these increments will not, in general, be equal; in fact, that they can be equal only when the three points of the curve are in the same straight line. If D is the point in which the line PQ meets the ordinate of R, then DR will be the difference between the two values of Δy, so that we shall have
$$DR = \Delta' y - \Delta y = \text{increment of } \Delta y.$$

Hence, again using the sign Δ to mark an increment, we shall have
$$DR = \Delta \Delta y \equiv \Delta^2 y, \qquad (b)$$
in which the exponent does not indicate a square, but merely the repetition of the symbol Δ.

THEOREM I. *When Δx becomes infinitesimal, $\Delta^2 y$ becomes an infinitesimal of the second order.*

For, if D be the point in which PQ produced cuts the ordinate $X_2 R$, we shall have, in the triangle QRD,
$$DR = QD \frac{\sin RQD}{\sin QRD} = \Delta^2 y. \qquad (b)$$

When Δx becomes an infinitesimal of the first order, so do both QD and the angle RQD, but the angle QRD will remain finite, because it will approach the angle QDX as its limit. Hence the expression will contain as a factor the product of two infinitesimals of the first order, and so will be an infinitesimal of the second order.

Since both the quantities QD and RQD depend upon Δx, we conclude that the ratio
$$\frac{\Delta^2 y}{\Delta x^2}$$
may remain finite when Δx becomes infinitesimal. In fact,

from the way we have formed these quantities, we have

$$\lim. \frac{\Delta^2 y}{\Delta x^2} = \lim. \frac{\Delta \frac{\Delta y}{\Delta x}}{\Delta x} = \frac{d \frac{dy}{dx}}{dx} = \phi''(x).$$

Hence—

Theorem II. *If we take two equal consecutive infinitesimal increments, $\equiv dx$, of the independent variable, then—*

1. *The difference between the corresponding infinitesimal increments of the function divided by dx^2 will approach a certain limit.*

2. *This limit is the derivative of the derivative of the function.*

Def. The derivative of the derivative is called the **second derivative**.

The derivative of the second derivative is called the **third derivative**, and so on indefinitely.

Notation. The successive derivatives of y with respect to x are written

$$\frac{dy}{dx}; \quad \frac{d^2y}{dx^2}; \quad \frac{d^3y}{dx^3}; \quad \text{etc.};$$

or
$$D_x y; \quad D_x^2 y; \quad D_x^3 y; \quad \text{etc.}$$

46. *Derivatives of any Order.* The results we have reached in the last article may be expressed thus: If we have an equation

$$y = \phi(x),$$

the first derivative is given by the equation

$$\frac{dy}{dx} = \phi'(x).$$

Then, by differentiating this equation, we have, by the last theorem,

$$\frac{d \cdot \frac{dy}{dx}}{dx} = \frac{d^2 y}{dx^2} = \phi''(x).$$

DERIVATIVES OF HIGHER ORDERS.

Again, taking the derivative, we have

$$\frac{d^3y}{dx^3} = \phi'''(x),$$

and we may continue the process indefinitely.

EXERCISES AND EXAMPLES.

1. To find the successive derivatives of ax^3.

$$\frac{dy}{dx} = 3ax^2;$$

$$\frac{d^2y}{dx^2} = 6ax;$$

$$\frac{d^3y}{dx^3} = 6a;$$

and all the higher derivatives will vanish.

Form the derivatives to the third, fourth or nth order of—

2. ax^4.
3. bx^{-1}.
4. $(a+x)^3$.
5. $(a-x)^3$.
6. $(a+x)^{-3}$.
7. $(a-x)^{-3}$.
8. $(a^2+x^2)^2$.
9. $2a^2x^2 + x^4$.
10. $a + bx + cx^2 + hx^3 + kx^4$.
11. $1 + x + x^2 + x^3 + x^4 + x^5 + \ldots + x^n$.
12. $1 - x + x^2 - x^3 + x^4 - x^5 + \ldots + (-1)^n x^n$.
13. $x^{\frac{1}{2}}$.
14. $x^{\frac{3}{2}}$.
15. $(a+x)^n$.
16. $(a+x)^{\frac{n}{2}}$.
17. If $y = e^x$, find $D_x^n y = a^x(\log a)^n$.
18. From $y = me^x$, find the nth derivative.
19. From $y = me^{hx}$ show that $D_x^n y = h^n y$.

Find the first three derivatives of the expressions:

20. z^x.
21. ax^x.
22. x^{ax}.
23. $\log x$.
24. $\log(a+x)$.
25. $m \log x$.
26. $\log(a-x)$.
27. $\log(a+mx)$.
28. $\log(a-mx)$.

29. Show that if $y = \sin x$, then $\frac{d^2y}{dx^2} = -y$.

$$\frac{d^4y}{dx^4} = y; \quad \frac{d^{n+4}y}{dx^{n+4}} = \frac{d^n y}{dx^n}.$$

30. Show that the same equations hold true if $y = \cos x$ or if $y = a \cos x + b \sin x$.

31. Find the law of formation of the successive derivatives of $\sin mx$ and $\cos mx$.

Especially, the $(n+4)$th derivative $= n$th der. $\times$ what?
$(n+2)$th derivative $= n$th der. $\times$ what?

32. Find the nth derivative of e^{mx}.

33. Find three derivatives of $e^{mz} \sin nz$.

34. If $u = y^y$, show that $\dfrac{d^2u}{dy^2} = (1 + \log y)\dfrac{du}{dy} + \dfrac{u}{y}$.

35. Find two derivatives of $u = \tan z$.

36. Find two derivatives of $u = \cos^2 z$.

37. Find two derivatives of $u = \sec^2 z$.

38. Find two derivatives of $u = \cos^2 z - \sin^2 z$.

39. Find two derivatives of $u = \cos 2z$.

40. Find two derivatives of $u = e^{-x^2}$.

41. Find two derivatives of $u = \sin^{(-1)} x$.

47. *Special Forms of Derivatives of Circular and Exponential Functions.* **Because**

$$\cos x = \sin(x + \tfrac{1}{2}\pi) \quad \text{and} \quad -\sin x = \cos(x + \tfrac{1}{2}\pi),$$

the derivatives of $\sin x$ and $\cos x$ may be written in the form

$$D_x \sin x = \sin(x + \tfrac{1}{2}\pi)$$
and
$$D_x \cos x = \cos(x + \tfrac{1}{2}\pi).$$

Hence, *the sine and cosine are such functions that their derivatives are formed by increasing their argument by $\tfrac{1}{2}\pi$.*

Differentiating by this rule n times in succession, we have

$$D_x^n \sin x = \frac{d^n \sin x}{dx^n} = \sin\left(x + \frac{n}{2}\pi\right);$$

$$D_x^n \cos x = \frac{d^n \cos x}{dx^n} = \cos\left(x + \frac{n}{2}\pi\right);$$

results which can be reduced to the forms found in Exercises 29 and 30 preceding.

48. Successive Derivatives of an Implicit Function. If the relation between y, the function, and x, the independent variable, is given in the implicit form

$$f(x, y) = 0,$$

then, putting u for this expression, we have found the first derivative to be

$$\frac{dy}{dx} = -\frac{\dfrac{du}{dx}}{\dfrac{du}{dy}}. \tag{a}$$

The values of both the numerator and denominator of the second member of this equation will be functions of x and y, which we may call $X_{\prime}$ and $Y_{\prime}$. We therefore write

$$\frac{dy}{dx} = -\frac{X_{\prime}}{Y_{\prime}}. \tag{b}$$

Differentiating this with respect to x, we shall have

$$\frac{d^2y}{dx^2} = \frac{-Y_{\prime}\dfrac{dX_{\prime}}{dx} + X_{\prime}\dfrac{dY_{\prime}}{dx}}{Y_{\prime}^2}. \tag{c}$$

$X_{\prime}$ and $Y_{\prime}$ being functions of both x and y, we have (§ 41)

$$\frac{dX_{\prime}}{dx} = \left(\frac{dX_{\prime}}{dx}\right) + \left(\frac{dX_{\prime}}{dy}\right)\frac{dy}{dx};$$

$$\frac{dY_{\prime}}{dx} = \left(\frac{dY_{\prime}}{dx}\right) + \left(\frac{dY_{\prime}}{dy}\right)\frac{dy}{dx}.$$

Substituting in these equations the values of $\dfrac{dy}{dx}$ from (b), and then substituting the results in (c), we shall have the required second derivative.

The process may then be repeated indefinitely, and thus the derivatives of any orders be found.

Example. Find the successive derivatives of y with respect to x from the equation

$$x^2 - xy + y^2 \equiv u = 0.$$

We have $\dfrac{du}{dx} = 2x - y$; $\dfrac{du}{dy} = -x + 2y$;

$$\dfrac{dy}{dx} = \dfrac{2x - y}{x - 2y}; \qquad (a')$$

which is a special case of (a) and (b), and where

$$X_{,} = 2x - y \quad \text{and} \quad Y_{,} = -x + 2y.$$

Differentiating the equation (a'), we have

$$\dfrac{d^2y}{dx^2} = \dfrac{(x - 2y)\dfrac{d(2x - y)}{dx} - (2x - y)\dfrac{d(x - 2y)}{dx}}{(x - 2y)^2}$$

$$= \dfrac{(x - 2y)\left(2 - \dfrac{dy}{dx}\right) + (2x - y)\left(2\dfrac{dy}{dx} - 1\right)}{(x - 2y)^2}.$$

Substituting the value of $\dfrac{dy}{dx}$ from (a'), we have

$$\dfrac{d^2y}{dx^2} = \dfrac{(x - 2y)(-3y) + 3x(2x - y)}{(x - 2y)^3}$$

$$= \dfrac{6(x^2 - xy + y^2)}{(x - 2y)^3} = \dfrac{6u}{(x - 2y)^3}.$$

EXERCISES.

Find by the above method the first two or three derivatives of v with respect to x, y or z, from the following equations:

1. $zv = a(v - z)$. Ans. $\dfrac{d^2v}{dz^2} = \dfrac{2(a + v)}{(a - z)^2}$.
2. $v^2y + vy^2 = a$.
3. $v^2 + vx + y^2 = b$.
4. $v(a - x)^2 + v^2(b + x) = c$.
5. $\log(v + z) + \log(v - z) = c$.
6. $\sin mv - \sin ny = h$.
7. $v(1 - a\cos z) = h$.
8. If $u - e \sin u = g$, show that

$$\dfrac{d^2u}{de\,dg} = \dfrac{1 - e}{(1 - e\cos u)^3}.$$

49. LEIBNITZ'S THEOREM. *To find the successive derivatives of a product in terms of the successive derivatives of its factors.*

Let $uv = p$ be the product of two functions of x. By successive differentiation we find

$$\frac{dp}{dx} = u\frac{dv}{dx} + v\frac{du}{dx};$$

$$\frac{d^2p}{dx^2} = u\frac{d^2v}{dx^2} + 2\frac{du}{dx}\frac{dv}{dx} + \frac{d^2u}{dx^2}v;$$

$$\frac{d^3p}{dx^3} = u\frac{d^3v}{dx^3} + 3\frac{du}{dx}\frac{d^2v}{dx^2} + 3\frac{d^2u}{dx^2}\frac{dv}{dx} + \frac{d^3u}{dx^3}v.$$

So far, the coefficients in the second member are those in the development of the powers of a binomial. To prove that this is true for the successive derivatives of every order, we note that each coefficient in any one equation is the sum of the corresponding coefficient plus the one to the left of it in the equation preceding. Now, let us have for any value of n

$$\frac{d^np}{dx^n} = u\frac{d^nv}{dx^n} + n\frac{du}{dx}\frac{d^{n-1}v}{dx^{n-1}} + \text{etc.;} \qquad (a)$$

the successive coefficients being

$$1; \quad n; \quad \left(\frac{n}{2}\right); \quad \left(\frac{n}{3}\right); \quad \text{etc.} \qquad (\text{Comp. § 6.})$$

Then, in the derivative of next higher order the coefficients will be

$$1; \quad n+1; \quad \left(\frac{n}{2}\right) + n \quad \text{or} \quad \left(\frac{n+1}{2}\right);$$

and, in general,

$$\left(\frac{n}{s}\right) + \left(\frac{n}{s-1}\right) \quad \text{or} \quad \left(\frac{n+1}{s}\right).$$

That is, $\frac{d^{n+1}p}{dx^{n+1}}$ is formed from (a) by writing $n+1$ for n. Hence, if the rule is true for n, it is also true for $n+1$. But it is true for $n = 3$; ∴ for $n = 4$, etc., indefinitely.

50. *Successive Derivatives with respect to Several Equicrescent Variables.* Studying the process of § 45, it will be seen that we supposed the successive increments of the independent variable to be equal to each other, and to remain equal as they became infinitesimal, while the increments of the functions were taken as variable. This supposition has been carried all through the subsequent articles.

Def. A variable whose successive increments are supposed equal is called an **equicrescent variable.**

We are now to consider the case of a function of several equicrescent variables.

If we have a function of two variables,
$$u = \phi(x, y),$$
the derivative of this function with respect to x will, in general, be a function of x and y. Let us write
$$\frac{du}{dx} = \phi_x(x, y).$$

Now, we may differentiate this equation with respect to y with a result of the form
$$\frac{d\frac{du}{dx}}{dy} = \phi_{x,y}(x, y).$$

Using a notation similar to that already adopted, we represent the first member of this equation in the form
$$\frac{d^2u}{dxdy}.$$

In the D-notation this is written
$$D^2_{x,y}u.$$

In either notation it is called "the second derivative of u with respect to x and y."

As an example: If we differentiate the function
$$u = y^2 \sin(mx - ny) \qquad (a)$$

with respect to x, and then differentiate the result with respect to y, we have

$$D_x u = \frac{du}{dx} = my^2 \cos(mx - ny);$$

$$D^2_{x, y} u = \frac{d^2 u}{dx\, dy} = 2my \cos(mx - ny) + mny^2 \sin(mx - ny).$$

51. We now have the following fundamental theorem:

$$\frac{d^2 u}{dx\, dy} = \frac{d^2 u}{dy\, dx};$$

or, in words,

The second derivative of a function with respect to two equicrescent variables is the same whether we differentiate in one order or the other.

Let $u = \phi(x, y)$ be the given function. Assigning to x the increment Δx, we have

$$\frac{\Delta u}{\Delta x} = \frac{\phi(x + \Delta x, y) - \phi(x, y)}{\Delta x}. \qquad (1)$$

In this equation assign to y the increment Δy, and call $\Delta \frac{\Delta u}{\Delta x}$ the corresponding increment of $\frac{\Delta u}{\Delta x}$. Then the equation will give

$$\frac{\Delta u}{\Delta x} + \Delta \frac{\Delta u}{\Delta x} = \frac{\phi(x + \Delta x, y + \Delta y) - \phi(x, y + \Delta y)}{\Delta x}.$$

Subtracting (1) and dividing the difference by Δy, we have

$$\frac{\Delta \frac{\Delta u}{\Delta x}}{\Delta y} = \frac{\phi(x + \Delta x, y + \Delta y) - \phi(x, y + \Delta y) - \phi(x + \Delta x, y) + \phi(x, y)}{\Delta x\, \Delta y}.$$

The second member of this equation is symmetrical with respect to x and y, and so remains unchanged when we interchange these symbols. Hence we have

$$\frac{\Delta \frac{\Delta u}{\Delta x}}{\Delta y} = \frac{\Delta \frac{\Delta u}{\Delta y}}{\Delta x}$$

for all values of Δx and Δy, and therefore for infinitesimal values of those increments. Thus

$$\frac{d\frac{du}{dx}}{dy} = \frac{d \cdot \frac{du}{dy}}{dx},$$

or $$D^2_{x,y} u = D^2_{y,x} u,$$

as was to be proved.

As an example, let us find the second derivative of (a) in the reverse order. We have

$$\frac{du}{dy} = 2y \sin(mx - ny) - ny^2 \cos(mx - ny);$$

$$\frac{d^2 u}{dy\,dx} = 2my \cos(mx - ny) + mny^2 \sin(mx - ny);$$

the same value as before.

COROLLARY. *The result of taking any number of successive derivatives of a function of any number of variables is independent of the order in which we perform the differentiations.*

For, by repeated interchanges of two successive differentiations, we can change the whole set of differentiations from one order to any other order.

If we have l differentiations with respect to x, m with respect to y, n with respect to z, etc., and use the D-notation, we express the result in the form

$$D_x^l D_y^m D_z^n \ldots \phi.$$

Here the symbol D_y^m means $D_y D_y$, etc., m times.

In the usual notation the same operation is expressed in the form

$$\frac{d^{l+m+n+\cdots} \phi}{dx^l dy^m dz^n \cdots}.$$

The corollary asserts that, using the D-notation, we may permute at pleasure the symbols D_x^l, D_y^m, D_z^n, etc., without changing the result of the differentiations.

EXERCISES.

Verify the theorem $D_x D_y u = D_y D_x u$ in the following cases:

1. $u = x \sin y + y \sin x$.   2. $u = x^y$.
3. $u = x \log y$.
4. $u = a \sin(x+y) - b \sin(x-y)$.

Differentiate each of the following functions once with respect to z, twice with respect to y, and three times with respect to x, in two different orders, and compare the results.

5. $\dfrac{x^3 y^2}{a^2 - z^2}$.   6. $ax^4 y^3 z^2$.

7. $x \sin y + y \sin z + z \sin x$.   8. $\sin(lx + my + nz)$.

9. If $u = \dfrac{1}{\sqrt{(x^2 + y^2 + z^2)}}$,

show that $\quad \dfrac{d^2 u}{dx^2} + \dfrac{d^2 u}{dy^2} + \dfrac{d^2 u}{dz^2} = 0.$

52. *Notation for Powers of a Differential or Derivative.* Such an expression as du^2 may be ambiguous unless defined. It may mean either

        Differential of square of u; i.e., $d(u^2)$;

or   Square of differential of u; i.e., $(du)^2$.

To avoid ambiguity, the expression as it stands is always supposed to have the latter meaning. To express the differential of the square of u we may write either

$$d \cdot u^2 \quad \text{or} \quad d(u^2),$$

of which the first form is the easier to use.

The square of the derivative $\dfrac{du}{dx}$ may be written either

$$\left(\dfrac{du}{dx}\right)^2 \quad \text{or} \quad \dfrac{du^2}{dx^2}.$$

CHAPTER VII.

SPECIAL CASES OF SUCCESSIVE DERIVATIVES.

53. *Successive Derivatives of a Power of a Derivative.* Let us have to differentiate the derivative

$$\left(\frac{du}{dx}\right)^2$$

with respect to x.

In such operations the D-notation will be found most convenient.

Applying the rule for differentiating a square, the result is

$$\frac{d\cdot\left(\frac{du}{dx}\right)^2}{dx} = 2\frac{du}{dx}\frac{d\cdot\frac{du}{dx}}{dx} = 2\frac{du}{dx}\frac{d^2u}{dx^2},$$

or, in the D-notation,

$$D_x(D_xu)^2 = 2D_xu D_x^2u.$$

In the same way, we find

$$\frac{d\cdot(D_xu)^n}{dx} = n\left(\frac{du}{dx}\right)^{n-1}\frac{d^2u}{dx^2} = n(D_xu)^{n-1}D_x^2u;$$

$$\frac{d\cdot(D_xu)^n}{dy} = n\left(\frac{du}{dx}\right)^{n-1}\frac{d^2u}{dxdy} = n(D_xu)^{n-1}D_{x,y}^2u;$$

$$\frac{d\cdot(D_x^2u)^2}{dx} = 2\frac{d^2u}{dx^2}\frac{d^3u}{dx^3} = 2D_x^2u D_x^3u.$$

SPECIAL CASES OF SUCCESSIVE DERIVATIVES. 87

EXERCISES.

Write the derivatives with respect to x of the following expressions, y being independent of x when it is written as an equicrescent variable:

1. $\left(\dfrac{dy}{dx}\right)^3$.
2. $\left(\dfrac{du}{dy}\right)^2$.
3. $\left(\dfrac{du}{dy}\right)^3$.
4. $y\left(\dfrac{dy}{dx}\right)^2$.
5. $y\left(\dfrac{dy}{dx}\right)^3$.
6. $y\left(\dfrac{du}{dy}\right)^3$.
7. $\left(\dfrac{d^2u}{dy^2}\right)^2$.
8. $\left(\dfrac{d^2u}{dx^2}\right)^3$.
9. $\left(\dfrac{d^2u}{dxdy}\right)^2$.
10. $\dfrac{du}{dx}\dfrac{du}{dy}$.
11. $\left(\dfrac{du}{dx}\right)^2\dfrac{du}{dy}$.
12. $\left(\dfrac{du}{dx}\right)^2\left(\dfrac{du}{dy}\right)^3$.
13. $\left(\dfrac{dy}{dx}\right)^2\left(\dfrac{du}{dz}\right)^3$.
14. $\left(\dfrac{du}{dx}\right)^2\left(\dfrac{d^2u}{dx^2}\right)^3$.
15. $\left(\dfrac{d^2u}{dy^2}\right)^2\left(\dfrac{d^3u}{dy^3}\right)^3$.
16. $\left(\dfrac{du}{dx}\right)^n\left(\dfrac{d^2u}{dx^2}\right)^m$.
17. $\left(\dfrac{d^3u}{dx^3}\right)^n$.
18. $\left(\dfrac{d^4u}{dx^4}\right)^n$.
19. $\left(\dfrac{d^2y}{dx^2}\right)^2\left(\dfrac{d^3y}{dx^3}\right)^3$.
20. $\dfrac{du}{dx}\dfrac{dv}{dy} - \dfrac{du}{dy}\dfrac{dv}{dx}$.
21. $\left(\dfrac{d^nu}{dy^n}\right)^n$.

54. *Derivatives of Functions of Functions.* Let us have, as in § 40,

$$u = f \cdot (\psi), \qquad (1)$$

where ψ is a given function of x. It is required to find the successive derivatives of u with respect to x. We may evidently reach this result by substituting in (1) for ψ its expression in terms of x, and then differentiating the result by methods already found.

But what we now wish to do is to find expressions for the successive derivatives without making this substitution. To do this, assign to x the infinitesimal increment dx. The resulting infinitesimal increment in ψ will be

$$d\psi = \dfrac{d\psi}{dx}dx.$$

This, again, will give u the increment

$$du = \frac{du}{d\psi}d\psi,$$

or, by substituting for $d\psi$ its value, and passing to the derivative,

$$\frac{du}{dx} = \frac{du}{d\psi}\frac{d\psi}{dx}. \qquad (2)$$

This is a particular case of the result already obtained in §40. The second member of (2) is a product of two factors. The first of these factors is formed by differentiating a function of ψ with respect to ψ; and is therefore another (derived) function of ψ; while the second is, for the same reason, a function of x.

Differentiating (2) with respect to x by the rule for a product, we have*

$$\frac{d^2u}{dx^2} = \frac{d\psi}{dx}\frac{d\frac{du}{d\psi}}{dx} + \frac{du}{d\psi}\frac{d^2\psi}{dx^2}. \qquad (3)$$

Now, because $\frac{du}{d\psi}$ is a function of ψ, its derivative with respect to x is to be obtained in the same way as that of u. If we put, for the moment,

$$u' = \frac{du}{d\psi} = f'(\psi),$$

we have, as in (2),

$$\frac{du'}{dx} = \frac{du'}{d\psi}\frac{d\psi}{dx} = \frac{d^2u}{d\psi^2}\frac{d\psi}{dx};$$

---

* The student should note that the expression $\dfrac{d\frac{du}{d\psi}}{dx}$ cannot be put in the form $\dfrac{d^2u}{d\psi dx}$, because the latter form presupposes that ψ and x are two independent variables, which is here not the case. In fact, u does not contain x except in ψ.

SPECIAL CASES OF SUCCESSIVE DERIVATIVES. 89

and hence, by substitution in (3),

$$\frac{d^2u}{dx^2} = \frac{d^2u}{d\psi^2}\left(\frac{d\psi}{dx}\right)^2 + \frac{du}{d\psi}\frac{d^2\psi}{dx^2}; \qquad (4)$$

which is the required expression for the second derivative.

From this we may form the third and higher derivatives by again applying the general rule embodied in (2), namely:

If ψ is a function of x, we find the derivative of any function, u, of ψ by differentiating u with respect to ψ, and multiplying the resulting derivative by $\dfrac{d\psi}{dx}$.

From the equation (4) we have

$$\frac{d^3u}{dx^3} = \left(\frac{d\psi}{dx}\right)^2 \frac{d \cdot \dfrac{d^2u}{d\psi^2}}{dx} + 2\frac{d^2u}{d\psi^2}\frac{d\psi}{dx}\frac{d^2\psi}{dx^2}$$

$$+ \frac{d^2\psi}{dx^2}\frac{d \cdot \dfrac{du}{d\psi}}{dx} + \frac{du}{d\psi}\frac{d^3\psi}{dx^3}.$$

By the rule just given, we have

$$\frac{d \cdot \dfrac{d^2u}{d\psi^2}}{dx} = \frac{d^3u}{d\psi^3}\frac{d\psi}{dx};$$

$$\frac{d \cdot \dfrac{du}{d\psi}}{dx} = \frac{d^2u}{d\psi^2}\frac{d\psi}{dx}.$$

Hence, by substitution and aggregation of like terms,

$$\frac{d^3u}{dx^3} = \frac{d^3u}{d\psi^3}\left(\frac{d\psi}{dx}\right)^3 + 3\frac{d^2u}{d\psi^2}\frac{d^2\psi}{dx^2}\frac{d\psi}{dx} + \frac{du}{d\psi}\frac{d^3\psi}{dx^3}. \qquad (5)$$

Repeating the process, we shall find

$$\frac{d^4u}{dx^4} = \frac{d^4u}{d\psi^4}\left(\frac{d\psi}{dx}\right)^4 + 6\frac{d^3u}{d\psi^3}\frac{d^2\psi}{dx^2}\left(\frac{d\psi}{dx}\right)^2$$

$$+ \frac{d^2u}{d\psi^2}\left[4\frac{d^3\psi}{dx^3}\frac{d\psi}{dx} + 3\left(\frac{d^2\psi}{dx^2}\right)^2\right] + \frac{du}{d\psi}\frac{d^4\psi}{dx^4}. \qquad (6)$$

EXAMPLE. Let us take the case of
$$u = \sin \psi,$$
ψ being any function whatever of x. We may then form the successive derivatives as follows:

$$\frac{du}{dx} = \frac{du}{d\psi}\frac{d\psi}{dx} = \cos\psi \frac{d\psi}{dx};$$

$$\frac{d^2u}{dx^2} = -\sin\psi\left(\frac{d\psi}{dx}\right)^2 + \cos\psi\frac{d^2\psi}{dx^2};$$

$$\frac{d^3u}{dx^3} = -\cos\psi\left(\frac{d\psi}{dx}\right)^3 - 2\sin\psi\frac{d\psi}{dx}\frac{d^2\psi}{dx^2}$$
$$\quad - \sin\psi\frac{d\psi}{dx}\frac{d^2\psi}{dx^2} + \cos\psi\frac{d^3\psi}{dx^3}$$
$$= -\cos\psi\left(\frac{d\psi}{dx}\right)^3 - 3\sin\psi\frac{d\psi}{dx}\frac{d^2\psi}{dx^2} + \cos\psi\,\frac{d^3\psi}{dx^3}.$$

EXERCISES.

Putting $\phi \equiv$ a function of x, find the first three derivatives of the following functions of ϕ with respect to x:

1. $u = \cos\phi$.
2. $u = \phi^2$.
3. $u = \phi^3$.
4. $u = \phi^n$.
5. $u = \log\phi$.
6. $u = e^\phi$.
7. $u = \sin 2\phi$.
8. $u = \cos 2\phi$.

55. *Change of the Equicrescent Variable.* Let the relation between y and x be expressed in the form

$$x = \phi(y), \qquad (1)$$

and let it be required to find the successive derivatives of y with respect to x, regarding the latter as the equicrescent. We may do this by solving (1) with respect to y, and then differentiating with respect to x in the usual way.

But the method of the last article will enable us to express the required successive derivatives of y with respect to x in terms of those of x with respect to y, which we can obtain from (1). By differentiating (1) as often as we please, we have results of the form

$$\begin{rcases} D_y x = \phi' y \equiv x'; \\ D_y^2 x = \phi'' y \equiv x''; \\ D_y^3 x = \phi''' y \equiv x'''. \end{rcases} \quad (2)$$

etc. etc.

x', x'', x''', etc., thus representing functions of y.

From § 37, Cor., we have

$$\frac{dy}{dx} = \frac{1}{D_y x} = \frac{1}{x'}. \quad (3)$$

To obtain the second derivative, we have to differentiate x', a function of y, with respect to x (§ 54). Thus

$$\frac{d^2 y}{dx^2} = -\frac{1}{x'^2} \cdot \frac{dx'}{dy} \cdot \frac{dy}{dx}.$$

From (2), $\quad \dfrac{dx'}{dy} = \dfrac{d^2 x}{dy^2} = x''.$

From this equation and (3) we have

$$\frac{d^2 y}{dx^2} = -\frac{x''}{x'^3} = -\frac{\dfrac{d^2 x}{dy^2}}{\left(\dfrac{dx}{dy}\right)^3}. \quad (4)$$

Differentiating again, we find

$$\frac{d^3 y}{dx^3} = \left(\frac{3x''}{x'^4} \cdot \frac{dx'}{dy} - \frac{1}{x'^3} \cdot \frac{dx''}{dy}\right) \frac{dy}{dx}$$

$$= \frac{3x''^2 - x' x'''}{x'^5} = \frac{3\left(\dfrac{d^2 x}{dy^2}\right)^2 - \dfrac{dx}{dy} \cdot \dfrac{d^3 x}{dy^3}}{\left(\dfrac{dx}{dy}\right)^5}.$$

The above process may be carried on to any extent. But many students will appreciate the following more elegant method of obtaining the required derivatives.

Imagine that we have solved the equation (1) so as to obtain a result in the form

$$y = F(x). \quad (5)$$

If in this equation we substitute for x its value (1), we shall have a result in the form

$$y = F(\phi y), \qquad (6)$$

which, of course, will really be an identity.

But we may still differentiate (5) with respect to y, regarding x as a function of y given by (1), by the method of §§ 40 and 54. Thus we shall have

$$\frac{dy}{dy} = \frac{dy}{dx}\frac{dx}{dy}; \qquad (\S\ 54,\ \text{Eq. 2.})$$

$$\frac{d^2y}{dy^2} = \frac{d^2y}{dx^2}\left(\frac{dx}{dy}\right)^2 + \frac{dy}{dx}\frac{d^2x}{dy^2}; \qquad (\S\ 54,\ \text{Eq. 4.})$$

$$\frac{d^3y}{dy^3} = \frac{d^3y}{dx^3}\left(\frac{dx}{dy}\right)^3 + 3\frac{d^2y}{dx^2}\frac{d^2x}{dy^2}\frac{dx}{dy} + \frac{dy}{dx}\frac{d^3x}{dy^3}.$$

etc.   etc.   etc.   etc.

But from the identity (6) $y = y$, which is obtained from (5), we have

$$\frac{dy}{dy} = 1;\quad \frac{d^2y}{dy^2} = 0;\quad \frac{d^3y}{dy^3} = 0;\quad \text{etc.}$$

Therefore, substituting for the derivatives of x with respect to y the expressions x', x'', etc., in (4), we have the equations

$$x'\frac{dy}{dx} = 1;$$

$$x'^2\frac{d^2y}{dx^2} + x''\frac{dy}{dx} = 0;$$

$$x'^3\frac{d^3y}{dx^3} + 3x'x''\frac{d^2y}{dx^2} + x'''\frac{dy}{dx} = 0;$$

$$x'^4\frac{d^4y}{dx^4} + 6x'^2x''\frac{d^3y}{dx^3} + (4x'x''' + 3x''^2)\frac{d^2y}{dx^2} + x^{\mathrm{iv}}\frac{dy}{dx} = 0.$$

Solving these equations successively, we shall find the values of $\frac{dy}{dx}$, $\frac{d^2y}{dx^2}$, etc., already obtained.

56. *Case of Two Variables Connected by a Third.* The case is still to be considered in which the relation between x

SPECIAL CASES OF SUCCESSIVE DERIVATIVES. 93

and y is expressed in the form

$$y = \phi_1(u); \quad x = \phi_2(u). \tag{1}$$

From these equations it is required to find the successive derivative of y with respect to x.

The first derivative is given by the equation

$$\frac{dy}{dx} = \frac{\frac{dy}{du}}{\frac{dx}{du}} = \frac{D_u y}{D_u x}.$$

From the manner in which the second member of this equation is formed, it is an explicit function of u alone. Hence (§ 54) we obtain its derivative with respect to x by taking its derivative with respect to u, and multiplying by $\frac{du}{dx}$. Thus

$$\frac{d^2y}{dx^2} = \frac{\frac{dx}{du}\frac{d^2y}{du^2} - \frac{dy}{du}\frac{d^2x}{du^2}}{\left(\frac{dx}{du}\right)^2} \cdot \frac{du}{dx}$$

$$= \frac{\frac{dx}{du}\frac{d^2y}{du^2} - \frac{dy}{du}\frac{d^2x}{du^2}}{\left(\frac{dx}{du}\right)^3}.$$

This, again, being a function of u, further derivatives with respect to x may be obtained by a repetition of the process.

EXERCISES.

Find the second derivative of x with respect to y, and also of y with respect to x, when the relation of x and y is given by the following equations:

1. $x = a \cos u;$ $y = b \sin u.$
2. $x = a \cos 2u;$ $y = b \sin u.$
3. $x = a \cos 2u;$ $y = b(\cos u - \sin u).$
4. $x = u - e \sin u;$ $y = u + e \sin u.$
5. $x = e^u;$ $y = ue^{au}.$

6. Show that if

$$y = e^u \cos u, \quad \text{then} \quad \frac{d^2u}{dy^2} = \frac{2 \sin u}{e^{2u}(\cos u - \sin u)^3}.$$

7. Show that the nth derivative of $x^n + ax^{n-1} + bx^{n-2}$ is $n!$, n being a positive integer > 1.

8. Show that

$$D_x^{\,3}(u^3) = 3u^2 D_x^{\,3}u + 18u D_x u D_x^{\,2}u + 6(D_x u)^3.$$

9. Show that if $v = u^n$, then

$$D_x^{\,3}v = nu^{n-1}D_x^{\,3}u + 3n(n-1)u^{n-2}D_x u D_x^{\,2}u \\ + n(n-1)(n-2)u^{n-3}(D_x u)^3.$$

10. If $u = a \cos mx + b \sin mx$, show that

$$D_x^{\,2}u + m^2 u = 0.$$

Then, by successively differentiating this result, show that, whatever the integer n,

$$D_x^{\,n+2}u + m^2 D_x^{\,n}u = 0;$$
$$D_x^{\,n+4}u - m^4 D_x^{\,n}u = 0.$$

11. If $u = e^x \cos x$ and $v = e^x \sin x$, then

$$D_x^{\,2}u = -2v \quad \text{and} \quad D_x^{\,2}v = 2u.$$

Also,
$$D_x^{\,4}v + 4v = 0;$$
$$D_x^{\,4}u + 4u = 0.$$

12. If $u = e^{nx} \cos mx$ and $v = e^{nx} \sin mx$, show that the successive derivatives of u and v may always be reduced to the form

$$D_x^{\,i}u = A_i u - B_i v; \quad D_x^{\,i}v = A_i v + B_i u, \qquad (a)$$

where A and B are functions of m and n. Also, find the values of A_1, A_2, B_1 and B_2, and show by differentiating (a) that

$$A_{i+1} = A_1 A_i - B_1 B_i; \quad B_{i+1} = B_1 A_i + A_1 B_i.$$

CHAPTER VIII.

DEVELOPMENTS IN SERIES.

57. A **series** is a succession of terms all of whose values are determined by any one rule.

A series is called

Finite when the number of its terms is limited;

Infinite when the number of its terms has no limit.

The sum of a finite series is the sum of all its terms.

The sum of an infinite series is the limit (if any) which the sum of its terms approaches as the number of terms added together is increased without limit.

When such a limit exists, the series is called **convergent**.

When it does not exist, the series is called **divergent**.

To **develop** a function means to find a series the limit of whose sum, if convergent, shall be equal to the function.

We may designate a series in the most general way, in the form

$$u_1 + u_2 + u_3 + \ldots + u_n + u_{n+1} + \ldots,$$

the nth terms being called u_n.

58. *Convergence and Divergence of Series.* No universal criterion has been found for determining whether any given series is convergent or divergent. There are, however, a great number of criteria applicable to a wide range of cases. Of these we mention the simplest.

I. *A series cannot be convergent unless, as n becomes infinite, the nth term approaches zero as its limit.*

For if, in such case, the limit of the terms is a finite quantity a, then each new term which we add will always change the sum of the series by at least α, and so that sum cannot approach a limit.

As an example, the sum of the series
$$1 - 1 + 1 - 1 + 1 - 1, \text{ etc., } ad\ infinitum,$$
will continually change from $+1$ to 0, and so can approach no limit, and so is divergent, by definition.

II. *A series all of whose terms are positive is divergent unless* $nu_n \doteq 0$ *when* $n \doteq \infty$.

To prove this, we have first to show that the harmonic series
$$\tfrac{1}{2} + \tfrac{1}{3} + \tfrac{1}{4} + \tfrac{1}{5} + \text{ etc., } ad\ infinitum,$$
is divergent. To do this we divide the terms of the series, after the first, into groups, the first group being the 2 terms $\tfrac{1}{3} + \tfrac{1}{4}$, the second group the following 4 terms, the third group the 8 terms next following, and, in general, the nth group the 2^n terms following the last preceding group. We shall then have an infinite number of groups, each greater than $\tfrac{1}{2}$.

Now, if, for all the terms of the series after the nth, we have
$$nu_n > \alpha \ (\alpha \text{ being any finite quantity}),$$
then
$$u_n > \frac{\alpha}{n},$$
and $u_m + u_{m+1} + \ldots > \alpha \left(\dfrac{1}{m} + \dfrac{1}{m+1} + \dfrac{1}{m+2} + \ldots \right)$.

Because the last factor of the second member of this equation increases to infinity, so does its product by α, which proves the theorem.

III. *If the terms of a series are alternately positive and negative, continually diminish, and approach zero as a limit, then the series is convergent.*

Let the series be
$$u_1 - u_2 + u_3 - u_4 + u_5 - \ldots.$$
Then, by hypothesis,
$$u_1 > u_2 > u_3 > u_4 > \ldots.$$

DEVELOPMENTS IN SERIES. 97

Let us put S_n for the sum of the first n terms of the series, n being any even integer, and S for the limit of the sum, if any there be. Then this limit may be expressed in either of the forms

$$S = S_n \;\;\; + (u_{n+1} - u_{n+2}) + (u_{n+3} - u_{n+4}) + \cdots$$

and

$$S = S_{n+1} - (u_{n+2} - u_{n+3}) - (u_{n+4} - u_{n+5}) - \cdots$$

Since all the differences in the parentheses are positive, by hypothesis it follows that, how many terms soever we take, the sum will always be greater than S_n and less than S_{n+1}. The difference of these quantities is u_{n+1}, which, by hypothesis, approaches zero as a limit. Since the two quantities S_n and S_{n+1} approach indefinitely near each other from opposite directions, they must each approach a limit S contained between them.

Graphically the demonstration may be shown to the eye thus: Let the line OS_6 represent the sum S_n, when $n = 6$,

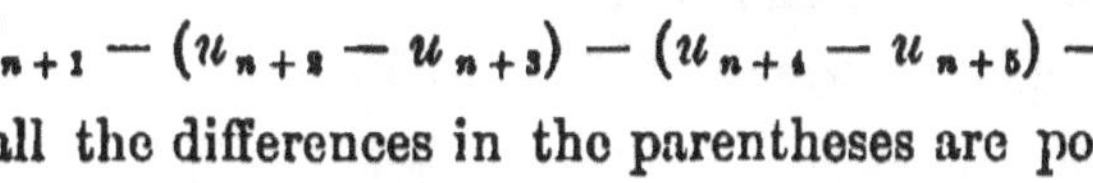

FIG. 11.

or any other even number; OS_7, the sum S_7, etc. Then every succeeding even sum is greater than that preceding, and every succeeding odd sum is less than that preceding, while the two approach each other indefinitely. Hence there must be some limit S which both approach.

An example of such a series is

$$1 - \frac{1}{3} + \frac{1}{5} - \frac{1}{7} + \frac{1}{9} - \text{etc.},$$

of which the nth term is $-\dfrac{(-1)^n}{2n-1}$. We shall hereafter see that the limit of the sum of this series is $\tfrac{1}{4}\pi$. If we divide the terms into pairs whose sums are negative, the series may be written

Pairing the terms so that the sum of each pair shall be positive, the series becomes

$$\frac{2}{3} + \frac{2}{5\cdot 7} + \frac{2}{9\cdot 11} + \frac{2}{13\cdot 15} + \text{etc.}$$

We may show by the preceding demonstration that these series approach the same limit.

IV. *If, after a certain finite number of terms, the ratio of two consecutive terms of a series is continually less than a certain quantity α, which is itself less than unity, then the series is convergent.*

Let the nth term be that after which the ratio is less than α. We then have

$$u_{n+1} < \alpha u_n;$$
$$u_{n+2} < \alpha u_{n+1} < \alpha^2 u_n;$$
$$u_{n+3} < \alpha u_{n+2} < \alpha^3 u_n;$$
$$\cdot \quad \cdot \quad \cdot \quad \cdot \quad \cdot \quad \cdot$$

Taking the sum of the members of these inequalities, we have

$$u_{n+1} + u_{n+2} + u_{n+3} + \ldots < (\alpha + \alpha^2 + \alpha^3 + \ldots)u_n.$$

But $\alpha + \alpha^2 + \alpha^3 + \ldots$ is an infinite geometrical progression whose limit when $\alpha < 1$ is $\dfrac{\alpha}{1-\alpha}$, a finite quantity.

Hence, putting S for the limit of the sum of the given series, we have

$$S < S_n + \frac{\alpha}{1-\alpha}u_n.$$

The second member of this inequality being a finite quantity which S can never reach, S must have some limit less than that quantity.

As an example, let us take the exponential series

$$e^x = 1 + x + \frac{x^2}{2!} + \frac{x^3}{3!} + \ldots$$

DEVELOPMENTS IN SERIES.

The ratio of the $(n+1)$st to the nth term is $\frac{x}{n}$. This ratio becomes less than unity when $n > x$, and it approaches zero as a limit. Hence the series is convergent for all values of x.

COROLLARY. *A series*

$$a_0 + a_1 x + a_2 x^2 + a_3 x^3 + \ldots$$

proceeding according to the powers of a variable, x, is convergent when $x < 1$, provided that the coefficients a_n do not increase indefinitely.

REMARKS.—(1) Note that, in applying the preceding rule, it does not suffice to show that the ratio of two consecutive terms is itself always less than unity. This is the case in the harmonic series, but the series is nevertheless divergent. The *limit* of the ratio must be less than unity.

(2) If the limit of the ratio in question is greater than unity, the series is of course divergent. Hence the only case in which Rule IV. leaves a doubt is that in which the ratio, being less than unity, approaches unity as a limit. But most of the series met with come into this class.

(3) The sum of a limited number of terms of a series gives no certain indication of its convergence or divergence. If we should compute the successive terms in the development of e^{-100} we should soon find ourselves dealing with numbers having thirty digits to the left of the decimal-point, and still increasing. But we know that if we should continue the computation far enough, say to 1000 terms, the positive and negative terms would so cancel each other that in writing the algebraic sum we should have 42 zeros to the right of the decimal-point.

On the other hand, if the whole human race, since the beginning of history, had occupied itself solely in computing the terms of the harmonic series, the sum it would have obtained up to the present time would have been less than 44. For 1000 million of people writing 5000 terms a day for 2 million of days would have written only 10^{19} terms. It is a theorem of the harmonic series, which we need not stop to demonstrate, that

$$S_n = \frac{1}{2} + \frac{1}{3} + \frac{1}{4} + \ldots + \frac{1}{n} < \text{Nap. log } n.$$

But $\quad$ Nap. log $10^{19} = \dfrac{\text{comm. log } 10^{19}}{0.4343\ldots} = \dfrac{19}{.4334\ldots} = 43.78$,

and yet the limit of the sum of the series is infinite.

59. Maclaurin's Theorem. This theorem gives a method of developing any function of a variable in a series proceeding according to the ascending powers of that variable.

If x represents the variable, and ϕ the function, the series to be investigated may be written in the form

$$\phi(x) = A_0 + A_1 x + A_2 x^2 + A_3 x^3 + \ldots; \qquad (1)$$

the series continuing to infinity unless ϕ is an entire function, in which case the two members are identical.

Whether the development (1) is or is not possible depends upon the form of the function ϕ. Most functions admit of being so developed; but special cases may arise in which the development is not possible. Moreover, the development will be illusory unless the series (1) is *convergent*. Commonly this series will be convergent for values of x below a certain magnitude, often *unity*, and divergent for values above that magnitude. What we shall now do is to assume the development possible, and show how the values of the coefficients A may be found.

Let us form the successive derivatives of the equation (1). We then have

$$\phi(x) = A_0 + A_1 x + A_2 x^2 + \text{etc.};$$
$$\frac{d\phi}{dx} = \phi'(x) = A_1 + 2A_2 x + 3A_3 x^2 + \ldots;$$
$$\frac{d^2\phi}{dx^2} = \phi''(x) = 1\cdot 2 A_2 + 2\cdot 3 A_3 x + 3\cdot 4 A_4 x^2 + \ldots;$$
$$\frac{d^3\phi}{dx^3} = \phi'''(x) = 1\cdot 2\cdot 3 A_3 + 2\cdot 3\cdot 4 A_4 x + \ldots;$$

$$\cdot \quad \cdot \quad \cdot \quad \cdot$$
$$\cdot \quad \cdot \quad \cdot \quad \cdot$$
$$\cdot \quad \cdot \quad \cdot \quad \cdot$$

$$\frac{d^n\phi}{dx^n} = \phi^{(n)}(x) = 1\cdot 2\cdot 3\cdot 4 \ldots n A_n + \text{etc.}$$

By hypothesis these equations are true for all values of x small enough to render the series convergent. Let us then put $x = 0$ in all of them. We then have

$\phi(0) = A_0;$ $\therefore A_0 = \phi(0).$
$\phi'(0) = A_1;$ $\therefore A_1 = \phi'(0).$
$\phi''(0) = 1\cdot 2 A_2;$ $\therefore A_2 = \dfrac{1}{1\cdot 2}\phi''(0).$
$\phi'''(0) = 1\cdot 2\cdot 3 A_3;$ $\therefore A_3 = \dfrac{1}{1\cdot 2\cdot 3}\phi'''(0).$

. . . .
. . . .

$\phi^{(n)}(0) = n! A_n;$ $\therefore A_n = \dfrac{1}{n!}\phi^{(n)}(0).$

By substituting these values in (1) we shall have the required development. Noticing that the symbolic forms $\phi'(0)$, $\phi''(0)$, etc., mean the values which the successive derivatives take when we put $x = 0$ after differentiation, we see that the coefficients are obtained by the following rule:

Form the successive derivatives of the given function.

After the derivatives are formed, suppose the variable to be zero in the original function and in each derivative.

Divide the quantities thus formed, in order, by 1; 1; 1·2; 1·2·3, etc., the divisor of the nth derivative being n!

The quotients will be the coefficients of the powers of the variable in the development, commencing with the zero power, or absolute term.

EXAMPLES AND EXERCISES.

1. To develop $(a + x)^n \equiv u$ in powers of x. We have

$u = (a + x)^n;$ $\therefore A_0 = a^n.$

$\dfrac{du}{dx} = n(a + x)^{n-1};$ $\therefore A_1 = na^{n-1}.$

$\dfrac{d^2u}{dx^2} = n(n-1)(a+x)^{n-2};$ $\therefore A_2 = \dfrac{n(n-1)}{1\cdot 2}a^{n-2}.$

.
.

$\dfrac{d^su}{dx^s} = n(n-1)\ldots(n-s+1)(a+x)^{n-s}.$

Thus the development is

$$(a+x)^n = a^n + na^{n-1}x + \binom{n}{2}a^{n-2}x^2 + \binom{n}{3}a^{n-3}x^3 + \ldots,$$

which is the binomial theorem.

2. Develop $(a-x)^n$ in the same way.
3. Develop $\log(1+x)$.

Here we shall have

$$\frac{du}{dx} = \frac{1}{1+x} = (1+x)^{-1};$$

$$\frac{d^2u}{dx^2} = -(1+x)^{-2};$$

$$\frac{d^3u}{dx^3} = 1\cdot 2(1+x)^{-3};$$

etc.   etc.

Noticing that $\log 1 = 0$, we shall find

$$\log(1+x) = x - \tfrac{1}{2}x^2 + \tfrac{1}{3}x^3 - \tfrac{1}{4}x^4 + \ldots.$$

4. Develop $\log(1-x)$.
5. Develop $\cos x$ and $\sin x$.

The successive derivatives of $\sin x$ are $\cos x$, $-\sin x$, $-\cos x$, $\sin x$, etc. By putting $x=0$, these become $1, 0, -1, 0, 1, 0$, etc. Thus we find

$$\cos x = 1 - \frac{x^2}{1\cdot 2} + \frac{x^4}{4!} - \frac{x^6}{6!} + \ldots;$$

$$\sin x = x - \frac{x^3}{3!} + \frac{x^5}{5!} - \frac{x^7}{7!} + \ldots.$$

6. Develop e^x, where e is the Naperian base.

$$Ans.\ e^x = 1 + x + \frac{x^2}{2!} + \frac{x^3}{3!} + \ldots.$$

7. Develop e^{-x}.
8. Show that

$$a^x = 1 + x\log a + \frac{(x\log a)^2}{1\cdot 2} + \frac{(x\log a)^3}{1\cdot 2\cdot 3} + \ldots.$$

9. Deduce $e^{\sin x} = 1 + x + \frac{x^2}{2} - \frac{3x^4}{4!} + \ldots.$

10. Develop $\sin(a+x)$ and $\cos(a+x)$ and thence, by comparing with the results of Ex. 5, prove the formulæ for the sine and cosine of the sum of two arcs. Find first

$$\sin(a+x) = \sin a \left(1 - \frac{x^2}{2!} + \ldots\right) + \cos a \left(x - \frac{x^3}{3!} + \ldots\right).$$

11. Develop $(1+e^x)^n$ and show that the result may be reduced to the form

$$2^n\left\{1 + \frac{n}{2}x + \frac{n^2+n}{2^2}\frac{x^2}{1.2} + \frac{n^3+3n^2}{2^3}\frac{x^3}{3!} + \ldots\right\}.$$

12. Develop $e^x \sin x$ and $e^x \cos x$ and deduce the results

$$e^x \sin x = x + 2\frac{x^2}{2!} + 2\frac{x^3}{3!} - 4\frac{x^5}{5!} - 8\frac{x^6}{6!} - \ldots$$

$$e^x \cos x = 1 + x - 2\frac{x^3}{3!} - 4\frac{x^4}{4!} - 4\frac{x^5}{5!} + \ldots$$

13. Develop $\cos^3 x$.

Begin by expressing $\cos^3 x$ in the form $\frac{1}{4}\cos 3x + \frac{3}{4}\cos x$.

14. Develop $\tan^{(-1)} x$.

This case affords us an example of how the process of development may often be greatly abbreviated. It has been shown that

$$\frac{d\cdot\tan^{(-1)}x}{dx} = \frac{1}{1+x^2} = 1 - x^2 + x^4 - x^6 + \text{etc.} \quad (a)$$

Now assume

$$\tan^{(-1)}x = A + A_1 x + A_2 x^2 + \text{etc.}$$

This gives

$$\frac{d\cdot\tan^{(-1)}x}{dx} = A_1 + 2A_2 x + 3A_3 x^2 + \text{etc.} \quad (b)$$

Comparing (a) and (b), we have

$$A_1 = 1;\quad A_3 = -\tfrac{1}{3};\quad A_5 = \tfrac{1}{5};\quad A_7 = -\tfrac{1}{7};\quad \text{etc.}$$

and

$$A_2 = A_4 = A_6 \ldots = 0.$$

The value of A is evidently zero. Hence

$$\tan^{(-1)}x = x - \tfrac{1}{3}x^3 + \tfrac{1}{5}x^5 - \tfrac{1}{7}x^7 + \text{etc.} \quad (c)$$

15. Develop $\sin^{(-1)} x$.

Since $\dfrac{d \cdot \sin^{(-1)} x}{dx} = (1-x^2)^{-\frac{1}{2}}$,

we may develop the derivative and proceed as in the last example. We shall thus find

$$\sin^{(-1)} x = \frac{x}{1} + \frac{1}{2} \cdot \frac{x^3}{3} + \frac{1 \cdot 3}{2 \cdot 4} \cdot \frac{x^5}{5} + \frac{1 \cdot 3 \cdot 5}{2 \cdot 4 \cdot 6} \cdot \frac{x^7}{7} + \text{etc.}$$

60. *Ratio of the Circumference of a Circle to its Diameter.* The preceding development of $\tan^{(-1)} x$ affords a method of computing the number π with great ease. The series (c) could be used for this purpose, but the convergence would be very slow. Series converging more rapidly may be obtained by the following device:

Let α, α', α'', etc., be several arcs whose sum is $45° = \frac{1}{4}\pi$. We then have

$$\tan(\alpha + \alpha' + \alpha'' + \text{etc.}) = 1.$$

Let t, t', t'', etc., be the tangents of the arcs α, α', α'', etc.

If there are but two arcs, α and α', we then have, by the addition theorem for tangents,

$$\frac{t+t'}{1-tt'} = 1; \quad \text{or} \quad t + t' = 1 - tt'.$$

If there are three arcs, α, α', and α'', we replace t' by $\dfrac{t'+t''}{1-t't''}$ in the last expression, and thus get

$$t + t' + t'' - tt't'' = 1 - tt' - t't'' - tt''.$$

We now have to find fractional values of t, t' and t'' of the form $\dfrac{1}{m}$, m being an integer, which will satisfy one of these equations. Unity is chosen as the numerator because the powers of the fraction are then more easily computed. The simplest fractions which satisfy the last equation are

$$t = \frac{1}{2}; \quad t' = \frac{1}{5}; \quad t'' = \frac{1}{8}.$$

We then have, from the development of $\tan^{(-1)} t$, etc.,

$$\alpha = \frac{1}{2} - \frac{1}{3\cdot 2^3} + \frac{1}{5\cdot 2^5} - \ldots ;$$

$$\alpha' = \frac{1}{5} - \frac{1}{3\cdot 5^3} + \frac{1}{5\cdot 5^5} - \ldots ;$$

$$\alpha'' = \frac{1}{8} - \frac{1}{3\cdot 8^3} + \frac{1}{5\cdot 8^5} - \ldots ;$$

$$\frac{1}{4}\pi = \alpha + \alpha' + \alpha''.$$

These series were used by Dase in computing π to 200 decimals.

A combination yet more rapid in ordinary use is found by determining α and α' by the conditions

$$\tan \alpha = \frac{1}{5};$$

$$4\alpha - \alpha' = \frac{1}{4}\pi.$$

We then have
$$\tan 2\alpha = \frac{2t}{1-t^2} = \frac{5}{12};$$

$$\tan 4\alpha = \frac{120}{119};$$

and because $\alpha' = 4\alpha - \frac{1}{4}\pi = 4\alpha - 45°$, we have

$$\tan \alpha' = \frac{\tan 4\alpha - 1}{\tan 4\alpha + 1} = \frac{1}{239}.$$

Hence we may compute π thus:

$$\alpha = \frac{1}{5} - \frac{1}{3\cdot 5^3} + \frac{1}{5\cdot 5^5} - \frac{1}{7\cdot 5^7} + \ldots ;$$

$$\alpha' = \frac{1}{239} - \frac{1}{3\cdot 239^3} + \frac{1}{5\cdot 239^5} - \ldots ;$$

$$\frac{1}{4}\pi = 4\alpha - \alpha'.$$

Ten or eleven terms of the first series, with four of the second, will give π to 15 places of decimals.

61. In developing functions by Maclaurin's theorem we may often be able to express the derivatives of a certain order as functions of those of a lower order. The process of finding the higher derivatives may then be abbreviated by retaining the derivatives of lower orders in a symbolic form, so far as possible.

EXAMPLES.

1. Let us develop
$$u = \log(1 + \sin x) \equiv \phi(x).$$
We now have
$$\phi'(x) = \frac{\cos x}{1 + \sin x} = \frac{1 - \sin x}{\cos x} = \sec x - \tan x;$$
$$\phi''(x) = \sec x \tan x - \sec^2 x = -\sec x \phi'(x).$$

Now, in continuing the differentiation, we use the last of these forms instead of the middle one. Thus
$$\phi'''(x) = -\sec x \tan x\, \phi'(x) - \sec x \phi''(x)$$
$$= -\sec x \tan x\, \phi'(x) + \sec^2 x \phi'(x)$$
$$= -\phi'(x)\phi''(x).$$

We may now find the successive derivatives symbolically. Omitting the symbol x after ϕ, we have
$$\phi^{iv} = -\phi'\phi''' - \phi''^2;$$
$$\phi^{v} = -\phi'\phi^{iv} - 3\phi''\phi''';$$
$$\phi^{vi} = -\phi'\phi^{v} - 4\phi''\phi^{iv} - 3\phi'''^2.$$
etc. etc.

Supposing $x = 0$,

$\phi(0) = 0;$	$\phi^{iv}(0) = -2;$
$\phi'(0) = +1;$	$\phi^{v}(0) = +5;$
$\phi''(0) = -1;$	$\phi^{vi}(0) = -16;$
$\phi'''(0) = +1;$	etc. etc.

Hence
$$\log(1 + \sin x) = x - \frac{x^2}{2} + \frac{x^3}{6} - \frac{x^4}{12} + \frac{x^5}{24} - \frac{x^6}{45} + \cdots$$

DEVELOPMENTS IN SERIES.

2. To develop $u = \tan x$.

Let us write the equation in the implicit form
$$u \cos x - \sin x = 0.$$
Then, by differentiation and division by $\cos x$, we find
$$D_x u = 1 + u^2;$$
$$D_x^2 u = 2u D_x u = 2u + 2u^3;$$
$$D_x^3 u = 2u D_x^2 u + 2(D_x u)^2;$$
$$D_x^4 u = 2u D_x^3 u + 8 D_x u D_x^2 u + 6(D_x^2 u)^2.$$

Putting $u = 0$, we find the even derivatives to vanish and the odd ones to become 1, 2, 16, etc. Hence
$$\tan x = x + \tfrac{1}{3}x^3 + \tfrac{2}{15}x^5 + \ldots$$

3. To develop $u = \sec x$.

Differentiating the form $u \cos x - 1 = 0$, we find
$$D_x u \cos x - u \sin x = 0. \qquad (a)$$
The successive derivatives of this equation may each be written in the form
$$M \cos x - N \sin x = 0. \qquad (b)$$
For, if we differentiate this equation with respect to x, it becomes
$$(D_x M - N) \cos x - (M + D_x N) \sin x = 0.$$
Hence the derivative of (b) may be formed by putting
$$M' = D_x M - N; \quad N' = M + D_x N, \qquad (c)$$
and writing M' and N' instead of M and N in the equation.

In (a) we have
$$M = D_x u; \quad N = u.$$
Then, by successive substitution in (c),

$M' = D_x^2 u - u;$ $\qquad N' = 2 D_x u;$
$M'' = D_x^3 u - 3 D_x u;$ $\qquad N'' = 3 D_x^2 u - u;$
$M''' = D_x^4 u - 6 D_x^2 u + u;$ $\qquad N''' = 4 D_x^3 u - 4 D_x u;$
$M^{\text{iv}} = D_x^5 u - 10 D_x^3 u + 5 D_x u;$ $\qquad N^{\text{iv}} = 5 D_x^4 u - 10 D_x^2 u + u.$
$M^{\text{v}} = D_x^6 u - 15 D_x^4 u + 15 D_x^2 u - u;$

108   THE DIFFERENTIAL CALCULUS.

When $x = 0$, we have $\sin x = 0$, $\cos x = 1$, $u = 1$, and hence $M = M' = \ldots = 0$ in all the equations. Thus we find, for $x = 0$,

$$D_x^{\,2} u = u = 1;$$
$$D_x^{\,4} u = 6 - 1 = 5;$$
$$D_x^{\,6} u = 75 - 15 + 1 = 61;$$
etc.   etc.;

while the odd derivatives all vanish. Hence

$$\sec x = 1 + \frac{1}{2} x^2 + \frac{5}{4!} x^4 + \frac{61}{6!} x^6 + \ldots .$$

62. *Taylor's Theorem.* Taylor's theorem differs from Maclaurin's only in the form of stating the problem and expressing the solution. The problem is stated as follows:

Having assigned to a variable x an increment h, it is required to develop any function of $x + h$ in powers of h.

Solution. Let ϕ be the function to be developed, and let us put

$$\left. \begin{array}{l} u = \phi(x); \\ u' = \phi(x + h). \end{array} \right\} \quad (1)$$

Assume

$$u' = X_0 + X_1 h + X_2 h^2 + X_3 h^3 + \text{etc.} \ldots \quad (2)$$

where X_0, X_1, etc., are functions of x to be determined.

Then, by successive differentiation, we have

$$\left. \begin{array}{l} \dfrac{du'}{dh} = X_1 + 2X_2 h + 3X_3 h^2 + 4X_4 h^3 + \text{etc.}; \\[4pt] \dfrac{d^2 u'}{dh^2} = 2X_2 + 2\cdot 3 X_3 h + 3\cdot 4 X_4 h^2 + \text{etc.}; \\[4pt] \dfrac{d^3 u'}{dh^3} = 1\cdot 2\cdot 3 X_3 + 2\cdot 3\cdot 4 X_4 h + \text{etc.} \\[4pt] \text{etc.} \quad\quad \text{etc.} \quad\quad \text{etc.} \end{array} \right\} \quad (3)$$

We now modify these equations by the following lemma:

If we have a function of the sum only of several quantities, the derivatives of that function with respect to those quantities will be equal to each other.

DEVELOPMENTS IN SERIES.

For if in $f(x + h)$ we assign an increment Δh to x and to h separately, the results will be $f(x + h + \Delta h)$ and $f(x + \Delta h + h)$, which are equal.

It follows that we have

$$\frac{du'}{dh} = \frac{du'}{dx}.$$

Now these equal derivatives, like u' itself, are functions of $x + h$ alone, so the lemma may be applied to as many successive derivatives as we please, giving

$$\frac{d^2u'}{dh^2} = \frac{d^2u'}{dx^2};$$
$$\frac{d^3u'}{dh^3} = \frac{d^3u'}{dx^3};$$
etc.   etc.

Now let the derivatives with respect to x be substituted for those with respect to h in equations (3), and let us suppose h to become zero in equations (2) and (3). Then u' and its derivatives will reduce to u and its derivatives, and we shall get

$$X_0 = u;$$
$$X_1 = \frac{du}{dx};$$
$$X_2 = \frac{1}{1 \cdot 2}\frac{d^2u}{dx^2};$$
$$X_3 = \frac{1}{1 \cdot 2 \cdot 3}\frac{d^3u}{dx^3};$$
$$\cdot$$
$$\cdot$$
$$\cdot$$
$$X_n = \frac{1}{n!}\frac{d^nu}{dx^n}.$$

Then, by substitution in (2), we shall have, for the required development,

$$u' = u + \frac{du}{dx}\frac{h}{1} + \frac{d^2u}{dx^2}\frac{h^2}{1 \cdot 2} + \frac{d^3u}{dx^3}\frac{h^3}{1 \cdot 2 \cdot 3} + \text{etc.}$$

This formula is called *Taylor's Theorem*, after Brook Taylor, who first discovered it.

EXAMPLES AND EXERCISES.

1. Develop $(x + h)^n$.

We proceed as follows:

$$u = x^n;$$
$$\frac{du}{dx} = nx^{n-1};$$
$$\frac{d^2u}{dx^2} = n(n-1)x^{n-2};$$
$$\frac{d^3u}{dx^3} = n(n-1)(n-2)x^{n-3};$$

etc. etc.

By substitution in the general formula we find

$$(x+h)^n = x^n + \frac{n}{1}x^{n-1}h + \frac{n(n-1)}{1\cdot 2}x^{n-2}h^2$$
$$+ \frac{n(n-1)(n-2)}{1\cdot 2\cdot 3}x^{n-3}h^3 + \ldots$$

2. Develop the exponential function a^{x+h} in powers of h.

$$Ans.\ a^x\left(1 + \log a\frac{h}{1} + (\log a)^2\frac{h^2}{1\cdot 2} + \ldots\right).$$

3. $\sin(x+h)$.   4. $\cos(x+h)$.
5. $\sin(x-h)$.   6. $\cos(x-h)$.
7. $\log(x+h)$.   8. $\log(x-h)$.
9. $\log\dfrac{x+h}{x-h}$.   10. $\log \cos x$.
11. $\cos^2(x+h)$.   12. $\sin^2(x-h)$.
13. $\tan^{(-1)}(x+h)$.   14. $\sin^{(-1)}(x-h)$.

15. Deduce the general formula

$$f\left(\frac{x}{1+x}\right) = f(x) - \frac{x^2}{1+x}f'(x) + \frac{x^4}{(1+x)^2}\frac{f''(x)}{1\cdot 2} - \text{etc.}$$

16. Prove, by differentiation and applying the algebraic theorem that in two equal series the coefficients of like powers of the variables must be equal, that if we have

$$\log(a_0 + a_1x + a_2x^2 + \ldots) = b_0 + b_1x + b_2x^2 + \ldots,$$

DEVELOPMENTS IN SERIES. 111

then the coefficients a and b are connected by the relations

$$b_0 = \log a_0;$$
$$a_0 b_1 = a_1;$$
$$2a_0 b_2 + a_1 b_1 = 2a_2;$$
$$3a_0 b_3 + 2a_1 b_2 + a_2 b_1 = 3a_3;$$
etc.   etc.   etc.

17. Hence show that $\dfrac{1}{1-x}$ is the logarithm of the sum of an infinite series whose first terms are

$$e\left(1 + x + \frac{3x^2}{2} + \frac{13x^3}{6} + \frac{73x^4}{24} + \ldots\right).$$

63. *Identity of Taylor's and Maclaurin's Theorems.* These two theorems, though different in form, are identical in principle.

To see how Taylor's theorem flows from Maclaurin's, notice that h in the former corresponds to x in the latter. The derivatives with respect to x in Taylor's theorem are the same as the derivatives with respect to h, and if we suppose $h = 0$ after differentiation Taylor's form of development can be derived at once from Maclaurin's.

Conversely, Maclaurin's theorem may be regarded as a special case of Taylor's theorem, in which we take zero as the original value of the variable, and thus make the increment equal to the variable. That is, if we put $f(x)$ in the form

$$f(0 + x),$$

and then, using x for h, develop in powers of x by Taylor's theorem, we shall have Maclaurin's theorem.

64. *Cases of Failure of Taylor's and Maclaurin's Theorems.* In order that a development in powers of a variable may have a determinate value it is necessary that none of the coefficients in the development shall become infinite and that the developed series shall be convergent.

For example, cosec x cannot be developed in powers of x, because when $x = 0$ the cosecant and all its derivatives become infinite.

65. *Extension of Taylor's Theorem to Functions of Several Variables.* Let us have the function

$$u = f(x, y). \tag{1}$$

It is required to develop this function when x and y both receive increments.

Let us first assign to x the increment h, and suppose y to remain constant. We then have, by Taylor's theorem,

$$f(x+h, y) = u + \frac{du}{dx}\frac{h}{1} + \frac{d^2u}{dx^2}\frac{h^2}{2!} + \frac{d^3u}{dx^3}\frac{h^3}{3!} + \cdots, \tag{2}$$

in which u, $\frac{du}{dx}$, etc., are all functions of y.

Next, assign to y the increment k. The first member of (2) will become $f(x+h, y+k)$. Developing the coefficients in the second member in powers of k, the result will be:

u will be changed into

$$u + \frac{du}{dy}\frac{k}{1} + \frac{d^2u}{dy^2}\frac{k^2}{2!} + \frac{d^3u}{dy^3}\frac{k^3}{3!} + \cdots;$$

$\frac{du}{dx} \equiv D_x u$ will be changed into

$$D_x u + \frac{d \cdot D_x u}{dy}\frac{k}{1} + \frac{d^2 D_x u}{dy^2}\frac{k^2}{2!} + \cdots;$$

$\frac{d^2u}{dx^2} \equiv D_x^2 u$ will be changed into

$$D_x^2 u + \frac{d \cdot D_x^2 u}{dy}\frac{k}{1} + \frac{d^2 D_x^2 u}{dy^2}\frac{k^2}{2!} + \cdots;$$

etc.   etc.   etc.

Substituting these changed values of the coefficients in (2) it will become

$$f(x+h, y+k) = u + \frac{du}{dy}\frac{k}{1} + \frac{d^2u}{dy^2}\frac{k^2}{2!} + \frac{d^3u}{dy^3}\frac{k^3}{3!} + \cdots$$

$$+ \frac{du}{dx}\frac{h}{1} + \frac{d^2u}{dxdy}\frac{h}{1}\frac{k}{1} + \frac{d^3u}{dxdy^2}\frac{h}{1}\frac{k^2}{2!} + \cdots$$

$$+ \frac{d^2u}{dx^2}\frac{h^2}{2!} + \frac{d^3u}{dx^2 dy}\frac{h^2}{2!}\frac{k}{1} + \frac{d^4u}{dx^2 dy^2}\frac{h^2}{2!}\frac{k^2}{2!}$$

$$+ \frac{d^3u}{dx^3}\frac{h^3}{3!} + \cdots.$$

DEVELOPMENTS IN SERIES. 113

Thus the function is developed in *powers and products* of the increments h and k.

The law of the series will be seen most clearly by using the D-notation. For each pair of positive and integral values of m and n we shall have the term

$$D_x^m D_y^n u \times \frac{h^m}{m!} \cdot \frac{k^n}{n!}.$$

If we collect in one line the terms of the development which are of the same order in h and k, we shall have:

Order of Terms.

1st. $D_x u \dfrac{h}{1} + D_y u \dfrac{k}{1}.$

2d. $D_x^2 u \dfrac{h^2}{2!} + D_x D_y u \dfrac{h}{1} \cdot \dfrac{k}{1} + D_y^2 u \dfrac{k^2}{2!}.$

3d. $D_x^3 u \dfrac{h^3}{3!} + D_x^2 D_y u \dfrac{h^2}{2!} \dfrac{k}{1} + D_x D_y^2 u \dfrac{h}{1} \dfrac{k^2}{2!} + D_y^3 u \dfrac{k^3}{3!};$

. . .
. . .
. . .

rth. $D_x^r u \dfrac{h^r}{r!} + D_x^{r-1} D_y u \dfrac{h^{r-1}}{(r-1)!} \dfrac{k}{1} + \ldots$

EXERCISES.

1. Show that in the preceding development the terms of the rth order may be written in the form

$$\frac{1}{r!}\left\{ h^r D_x^r u + \binom{r}{1} h^{r-1} k D_x^{r-1} D_y u + \binom{r}{2} h^{r-2} k^2 D_x^{r-2} D_y^2 u + \ldots \right\},$$

$\binom{r}{1}$, $\binom{r}{2}$, etc., denoting the binomial coefficients as in § 5.

2. Extend the development to the case of three independent variables, and show that the terms to the second order inclusive will be as follows:

If
$$u = f(x, y, z),$$
then $f \cdot (x+h, y+k, z+l) = u$
$$+ D_x u \cdot h + D_y u \cdot k + D_z u \cdot l$$
$$+ D_x^2 u \cdot \frac{h^2}{2!} + D_y^2 u \cdot \frac{k^2}{2!} + D_z^2 u \cdot \frac{l^2}{2!} + D_x D_y u \cdot hk$$
$$+ D_x u D_z u \cdot hl + D_y u D_z u \cdot kl.$$

66. *Hyperbolic Functions.* The sine and cosine of an imaginary arc may be found as follows: In the developments for sin x and cos x, namely,

$$\sin x = x - \frac{x^3}{3!} + \frac{x^5}{5!} - \ldots,$$

$$\cos x = 1 - \frac{x^2}{2!} + \frac{x^4}{4!} - \ldots,$$

let us put yi for x. ($i \equiv \sqrt{-1}$). We thus have

$$\left. \begin{array}{l} \sin yi = i\left(y + \dfrac{y^3}{3!} + \dfrac{y^5}{5!} + \ldots \right); \\ \cos yi = 1 + \dfrac{y^2}{2!} + \dfrac{y^4}{4!} + \ldots \end{array} \right\} \quad (1)$$

We conclude:

The cosine of a purely imaginary arc is real and greater than unity, while its sine is purely imaginary.

We find from (1),

$$\cos yi + i \sin yi = 1 - y + \frac{y^2}{2!} - \text{etc.} = e^{-y};$$

$$\cos yi - i \sin yi = 1 + y + \frac{y^2}{2!} + \text{etc.} = e^{y};$$

and, by addition and subtraction,

$$\cos yi = \tfrac{1}{2}(e^{-y} + e^{y});$$
$$i \sin yi = \tfrac{1}{2}(e^{-y} - e^{y});$$
$$\sin yi = \tfrac{1}{2}i(e^{y} - e^{-y}).$$

The cosine of yi is called the **hyperbolic cosine** of y, and is written cosh y, the letter h meaning "hyperbolic."

DEVELOPMENTS IN SERIES. 115

The real factor in the sine of yi is called the **hyperbolic sine** of y, and is written $\sinh y$.

Thus the hyperbolic sine and cosine of a real quantity are real functions defined by the equations

$$\left.\begin{array}{l}\sinh y = \tfrac{1}{2}(e^y - e^{-y}); \\ \cosh y = \tfrac{1}{2}(e^y + e^{-y}).\end{array}\right\} \qquad (1)$$

By analogy, we introduce the additional function

$$\tanh y = \frac{e^y - e^{-y}}{e^y + e^{-y}}.$$

The differentiation of these expressions gives

$$\frac{d \sinh y}{dy} = \cosh y; \quad \frac{d \cosh y}{dy} = \sinh y; \qquad (2)$$

$$d \tanh y = \frac{dy}{\cosh^2 y}.$$

They also give the relations

$$\cosh^2 y - \sinh^2 y = 1. \qquad (3)$$

Inverse Hyperbolic Functions. When we form the inverse function, we may put

$$u \equiv \cosh y.$$

Then, solving the equation

$$e^y + e^{-y} = 2 \cosh y = 2u,$$

we find
$$e^y = u \pm \sqrt{u^2 - 1}.$$

Hence
$$y = \log(u \pm \sqrt{u^2 - 1}) = \cosh^{(-1)} u. \qquad (4)$$

In the same way, if we put

$$u \equiv \sinh y,$$

we find
$$y = \log(u \pm \sqrt{u^2 + 1}) = \sinh^{(-1)} u. \qquad (5)$$

From the equations (2) and (3) we find, for the derivatives of the inverse functions:

When $y = \cosh^{(-1)} u$, or $u = \cosh y$,

then
$$\frac{dy}{du} = \frac{1}{\sqrt{u^2 - 1}}. \tag{6}$$

When $y = \sinh^{(-1)} u$, or $u = \sinh y$,

then
$$\frac{dy}{du} = \frac{1}{\sqrt{u^2 + 1}}. \tag{7}$$

REMARK. The above functions are called *hyperbolic* because $\sinh y$ and $\cosh y$ may be represented by the co-ordinates of points on an equilateral hyperbola whose semi-axis is unity. The equation of such an hyperbola is

$$x^2 - y^2 = 1,$$

which is of the same form as (3).

EXERCISES.

1. By continuing the differentiation begun in (2) prove the following equations:

$$D_x^2 \sinh x = \sinh x;$$
$$D_x^2 \cosh x = \cosh x;$$
$$D_x^{n+2} \sinh x = \sinh x.$$
etc.   etc.

2. Develop $\sinh x$, as defined in (1), in powers of x by Maclaurin's theorem.

$$Ans. \; \sinh x = y + \frac{y^3}{3!} + \frac{y^5}{5!} + \cdots$$

3. Develop $\sinh (x + h)$ and $\cosh (x + h)$ by Taylor's theorem and deduce

$$\sinh (x+h) = \sinh x \left(1 + \frac{h^2}{2!} + \cdots \right) + \cosh x \left(x + \frac{x^3}{3!} + \cdots \right)$$
$$= \sinh x \cosh h + \cosh x \sinh h;$$
$$\cosh (x+h) = \cosh x \cosh h + \sinh x \sinh h.$$

CHAPTER IX.

MAXIMA AND MINIMA OF FUNCTIONS OF A SINGLE VARIABLE.

67. *Def.* A **maximum** value of a function is one which is greater than the values immediately preceding and following it.

A **minimum** value is one which is less than the values immediately preceding and following it.

REMARK. Since a maximum or minimum value does not mean the greatest or least possible value, a function may have several maxima or minima.

68. PROBLEM. *Having given a function*

$$y = \phi(x),$$

it is required to find those values of x for which y is a maximum or a minimum.

Let us assign to x the increments $+h$ and $-h$, and develop in powers of h. We shall then have

$$y' \equiv \phi(x-h) = y - \frac{dy}{dx}\frac{h}{1} + \frac{d^2y}{dx^2}\frac{h^2}{1\cdot 2} - \text{etc.};$$

$$y'' \equiv \phi(x+h) = y + \frac{dy}{dx}\frac{h}{1} + \frac{d^2y}{dx^2}\frac{h^2}{1\cdot 2} + \text{etc.}$$

In order that the value of $y = \phi(x)$ may be a minimum, it must, however small we suppose h, be less than either y' or y''. That is, the expressions

$$y' - y = -\frac{dy}{dx}\frac{h}{1} + \frac{d^2y}{dx^2}\frac{h^2}{1\cdot 2} - \text{etc.},$$

$$y'' - y = +\frac{dy}{dx}\frac{h}{1} + \frac{d^2y}{dx^2}\frac{h^2}{1\cdot 2} + \text{etc.},$$

must both be positive as h approaches zero. But if $\frac{dy}{dx}$ is finite, h may always be made so small that the terms in h^2 shall be less in absolute magnitude than those in h (§ 14), and the condition of a minimum cannot be satisfied. We must therefore have, as the first condition,

$$\frac{dy}{dx} \equiv \phi'(x) = 0. \tag{1}$$

By solving this equation with respect to x will be found a value of x called a *critical value*.

The same reasoning applies to the case of a maximum, so that the condition (1) is necessary to either a maximum or a minimum. Supposing it fulfilled, we have

$$y' - y = \frac{d^2y}{dx^2}\frac{h^2}{1\cdot 2} - \frac{d^3y}{dx^3}\frac{h^3}{1\cdot 2\cdot 3} + \text{etc.};$$

$$y'' - y = \frac{d^2y}{dx^2}\frac{h^2}{1\cdot 2} + \frac{d^3y}{dx^3}\frac{h^3}{1\cdot 2\cdot 3} + \text{etc.}$$

Since h^2 is positive, the algebraic sign of these quantities, as h approaches zero, will be the same as that of $\frac{d^2y}{dx^2}$.

When this second derivative is *positive* for the critical value of x, y, being *less* than y' or y'', will be a *minimum*.

When *negative*, y will be greater than either y' or y'', and so will be a *maximum*.

We therefore conclude:

Conditions of minimum: $\frac{dy}{dx} = 0$; $\frac{d^2y}{dx^2}$ *positive.*

Conditions of maximum: $\frac{dy}{dx} = 0$; $\frac{d^2y}{dx^2}$ *negative.*

We have, therefore, the rule:

Equate the first derivative of the function to zero. This equation will give one or more values of the independent variable, called critical values, and thence corresponding values of the function.

MAXIMA AND MINIMA.

Substitute the critical values in the expression for the second derivative. When the result is positive, the function is a minimum; when negative, a maximum.

Exceptional Cases. It may happen that the second derivative is zero for a critical value of x. We shall then have

$$y' - y = -\frac{d^3y}{dx^3}\frac{h^3}{3!} + \frac{d^4y}{dx^4}\frac{h^4}{4!} - \text{etc.};$$

$$y'' - y = \frac{d^3y}{dx^3}\frac{h^3}{3!} + \frac{d^4y}{dx^4}\frac{h^4}{4!} + \text{etc.};$$

and there can be neither a maximum nor a minimum unless $\frac{d^3y}{dx^3} = 0$. If this condition is fulfilled, y will be a maximum when the fourth derivative is negative; a minimum when it is positive.

Continuing the reasoning, we are led to the following extension of the rule:

Find the first derivative in order which does not vanish for a critical value of the independent variable. If this derivative is of an odd order, there is neither a maximum nor a minimum; if of an even order, there is a minimum when the derivative is positive, a maximum when it is negative.

The above reasoning may be illustrated by the graphic representation of the function. When the ordinate of the curve is a maximum or a minimum the tangent will be parallel to the axis of abscissas, and the angle which it makes with this axis will change from positive to negative at a point having a maximum ordinate, and from negative to positive at a point having a minimum ordinate.

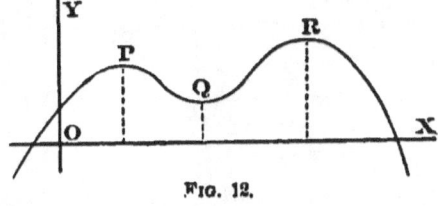

FIG. 12.

For example, in the figure a minimum ordinate occurs at the point Q, and maximum ordinates at P and R.

EXAMPLES AND EXERCISES.

1. Find the maximum and minimum values of the expression
$$y = 2x^3 + 3x^2 - 36x + 15.$$
By differentiation,
$$\frac{dy}{dx} = 6x^2 + 6x - 36;$$
$$\frac{d^2y}{dx^2} = 12x + 6.$$

Equating the first derivative to zero, we have the quadratic equation
$$x^2 + x - 6 = 0,$$
of which the roots are $x = 2$ and $x = -3$.

The values of $\frac{d^2x}{dx^2}$ are $+30$ and -30.

Hence $x = 2$ gives a minimum value of $y = -29$;
$\qquad x = -3$ gives a maximum value of $y = +95$.

Find the maximum and minimum values of the following functions:

2. $x^3 + 3x^2 - 24x + 9$.   3. $x^3 - 3x + 5$.

4. $y = \dfrac{x}{1 + x^2}$.   5. $y = \dfrac{x^2 - x + 1}{x^2 + x - 1}$.

6. $y = \dfrac{\log x}{x}$.   7. $y = \dfrac{\log x}{x^n}$.

8. $y = x^x$.   9. $y = \sin 2x - x$.

10. $y = (x + 1)(x - 2)^2$.   11. $y = (x - a^2)(x - b)^2$.

12. $y = \dfrac{(x + 3)^3}{(x + 2)^2}$.   13. $y = \dfrac{(x - a)(x - b)}{(x - p)(x - q)}$.

14. $y = \cos 2x$.   15. $y = \cos nx$.

16. $y = \sin 3x$.   17. $y = \sin nx$.

18. $y = \dfrac{x}{1 + x \tan x}$.   *Ans.* A maximum when $x = +\cos x$.
A minimum when $x = -\cos x$.

MAXIMA AND MINIMA. 121

19. $y = \sin^3 x \cos x$.   20. $y = \sin^3 x \cos x$.

21. $y = \dfrac{\sin x}{1 + \tan x}$.   22. $y = \dfrac{\cos x}{1 + \tan x}$.

23. The sum of two adjacent sides of a rectangle is equal to a fixed line a. Into what parts must a be divided that the rectangle may be a maximum?   *Ans.* Each part $= \tfrac{1}{2}a$.

<small>Note that the expression for the area is $x(a - x)$.</small>

24. Into what parts must a number be divided in order that the product of one part by the square of the other may be a maximum?   *Ans.* Into parts whose ratio is $1 : 2$.

<small>Note that if a be the number, the parts may be called x and $a - x$.</small>

25. Into what two parts must a number be divided in order that the product of the mth power of one part into the nth power of the other may be a maximum?

Ans. Into parts whose ratio is $m : n$.

26. Show that the quadratic function $ax^2 + bx + c$ can have but one critical value, and that it will depend upon the sign of the coefficient a whether that value is a maximum or a minimum.

27. A line is required to pass through a fixed point P, whose co-ordinates are a and b in the plane of a pair of rectangular axes OX and OY. What angle must the line make with the axis of X, that the area of the triangle XYO may be a minimum? Show also that P must bisect the segment XY.

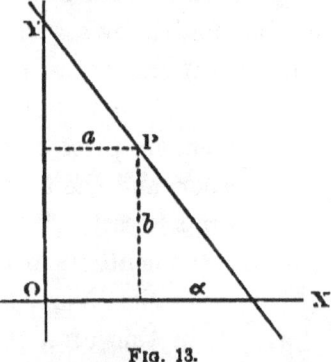

Fig. 13.

Express the intercepts which the line cuts off from the axes in terms of a, b and the variable angle α. The half product of these intercepts will be the area.

We shall thus find

$$2 \text{ Area} = (a + b \cot \alpha)(b + a \tan \alpha) = 2ab + a^2 \tan \alpha + \frac{b^2}{\tan \alpha}.$$

Then, taking $\tan \alpha \equiv t$ as the independent variable, we readily find, for the critical values of t and α,

$$t = \pm \frac{b}{a}, \quad \text{or} \quad a \sin \alpha = \pm b \cos \alpha.$$

It is then to be shown that both values of t give minima values of the area; that the one minimum area is $2ab$, and the other zero; that in the first case the line YX is bisected at P, and in the other case passes through O.

28. Show by the preceding figure that whatever be the angle XOY, the area of the triangle will be a minimum when the line turning on P is bisected at P.

The student should do this by drawing through P a line making a small angle with XPY. The increment of the area XOY will then be the difference of the two small triangles thus formed. Then let the small angle become infinitesimal, and show that the increment of the area XOY can become an infinitesimal of the second order only when $PX = PY$.

29. A carpenter has boards enough for a fence 40 feet in length, which is to form three sides of an enclosure bounded on the fourth by a wall already built. What are the sides and area of the largest enclosure he can build out of his material?   *Ans.* 10×20 feet $= 200$ square feet.

30. A square piece of tin is to have a square cut out from each corner, and the four projecting flaps are to be bent up so as to form a vessel. What must be the side of the part cut out that the contents of the vessel may be a maximum?

Ans. One sixth the side of the square.

31. If, in this case, the tin is a rectangle whose sides are $2a$ and $2b$, show that the side of the flap is

$$\tfrac{1}{2}(a + b - \sqrt{a^2 - ab + b^2}).$$

32. What is the form of the rectangle of greatest area which can be drawn in a semicircle?

Note that if r be the radius of the circle, and x the altitude of the rectangle, $\sqrt{r^2 - x^2}$ will be half the base of the rectangle.

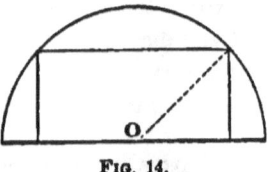

Fig. 14.

MAXIMA AND MINIMA. 123

69. *Case when the function which is to be a maximum or minimum is expressed as a function of two or more variables connected by equations of condition.*

The function which is to be a maximum or minimum may be expressed as a function of two variables, x and y, thus:

$$u = \phi(x, y). \qquad (1)$$

If x and y are independent of each other, the problem is different from that now treated.

If between them there exists some relation

$$f(x, y) = 0, \qquad (2)$$

we may, by solving this equation, express one in terms of the other, say y in terms of x. Then substituting this value of y in (1), u will be a function of x alone, which we may treat as before.

It may be, however, that the solution of the equation (2) will be long or troublesome. We may then avoid it by the method of § 41. From (1) we have

$$\frac{du}{dx} = \left(\frac{du}{dx}\right) + \left(\frac{du}{dy}\right)\frac{dy}{dx},$$

and from (2) we have, by the method of § 37,

$$\frac{dy}{dx} = -\frac{D_x f}{D_y f}.$$

Substituting this value in the preceding equation, we shall have the value of $\dfrac{du}{dx}$, which is to be equated to zero. The equation thus formed, combined with (2), will give the critical values of both x and y, and hence the maximum or minimum value of u.

EXAMPLES AND EXERCISES.

1. To find the form of that cylinder which has the maximum volume with a given extent of surface.

The total extent of surface includes the two ends and the convex cylindrical surface. If r be the radius of the base, and h the altitude, we shall have:

Area of base, πr^2.
Area of convex surface, $2\pi rh$.

Hence total surface $= 2\pi(r^2 + rh) = \text{const.} \equiv a$.   (a)

Also, volume $= \pi r^2 h$.   (b)

Putting u for the volume, we have, from (b),

$$\frac{du}{dr} = 2\pi rh + \pi r^2 \frac{dh}{dr}.$$

From (a) we find $\quad \dfrac{dh}{dr} = -\dfrac{h+2r}{r}.$

Whence $\quad \dfrac{du}{dr} = \pi rh - 2\pi r^2.$

Equating this to zero, we find that the altitude of the cylinder must be equal to the diameter of its base.

2. Find the shape of the largest cylindrical tin mug which can be made with a given weight of tin.

This problem differs from the preceding one in that the top is supposed to be open, so that the total surface is that of the base and convex portion.

Ans. Altitude $=$ radius of bottom.

3. Find the maximum rectangle which can be inscribed in a given ellipse.

If the equation of the ellipse is $b^2x^2 + a^2y^2 = a^2b^2$, the sides of the rectangle are $2x$ and $2y$. Hence the function to be a maximum is $4xy$, subject to the condition expressed by the equation of the ellipse. This condition gives

$$\frac{dy}{dx} = -\frac{b^2x}{a^2y}.$$

We shall find the rectangle to be a maximum when its sides are proportional to the corresponding axes of the ellipse; each side is then equal to the corresponding axis divided by $\sqrt{2}$.

4. Find the maximum rectangle which can be inscribed in the segment of a parabola whose semi-parameter is p, cut off by a double ordinate whose distance, OX, from the vertex is a. Show also that the ratio of its area to that of the circumscribed rectangle is constant and equal to

$$2 : \sqrt{27}.$$

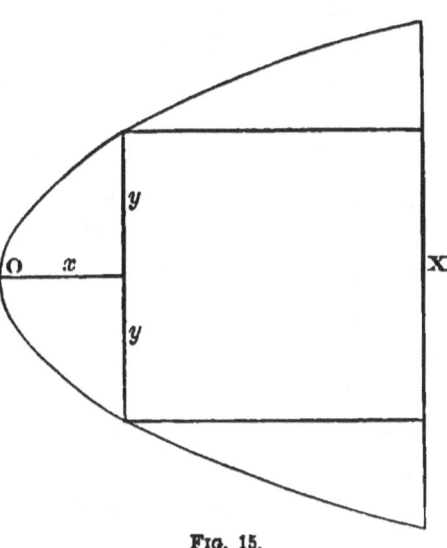

Fig. 15.

By taking x and y as in the figure, $a - x$ will be the base of the rectangle, and we shall have $2y$ for its altitude. Hence its area will be $2y(a - x)$, while x and y will be connected by the equation of the parabola, $y^2 = 2px$.

5. Find the cone of maximum volume which shall have a given extent of conical surface.

Ans. Alt. = radius of base $\times \sqrt{2}$.

6. Find the volume of the maximum cylinder which can be inscribed in a given right cone, and show that the ratio of its volume to that of the cone is $4 : 9$.

7. Find the cylinder of maximum cylindrical surface which can be inscribed in a right cone.

Ans. Alt. of cylinder = $\frac{1}{2}$ alt. of cone.

8. Find the maximum cone which can be inscribed in a given sphere.

If we make a central section of the sphere through the vertex of the cone, the base and slant height of the cone will be the base and equal sides of an isosceles triangle inscribed in the circular section. Thus the equation between the base and altitude of the cone can be obtained.

Ans. Alt. = $\frac{4}{3}$ radius of sphere.

9. Find the maximum cylinder which can be inscribed in an ellipsoid of revolution.

Ans. Alt. = $\dfrac{2}{\sqrt{3}}$ of axis of revolution.

10. Find the cone of maximum conical surface which can be inscribed in a given sphere.

11. Of all cones having the same slant height, which has the maximum volume?

12. A boatman 3 miles from the shore wishes, by rowing to the shore and then walking, to reach in the shortest time a point on the beach 5 miles from the nearest point of the shore. If he can pull 4 miles an hour and walk 5 miles an hour, to what point of the beach should he direct his course?

Ans. 4 miles from the nearest point of the shore.

Express the whole time required in terms of the distance x of his point of landing from the nearest point of the shore.

13. Find the maximum cone which can be inscribed in a paraboloid of revolution, the vertex of the cone being at the centre of the base of the paraboloid.

Ans. Alt. = $\frac{1}{3}$ alt. of paraboloid.

14. Find the maximum cylinder which can be described in a paraboloid of revolution.

15. Find the rectangle of maximum perimeter which can be inscribed in an ellipse.

16. On the axis of the parabola $y^2 = 2px$ a point is taken at distance a from the vertex. Find the abscissa of the nearest point of the curve.

Begin by expressing the square of the distance from the fixed point to the variable point (x, y) on the parabola.

17. Determine the cone of minimum volume which can be circumscribed around a given sphere.

18. Determine the cone of minimum conical surface which can be circumscribed around a given sphere.

19. Find that point on the line joining the centres of two circles from which the greatest length of the combined circumferences will be visible.

20. Find that point on the line joining the centres of two spheres of radii a and b respectively from which the greatest extent of spherical surface will be visible.

Ans. The point dividing the central line in the ratio $a^{\frac{1}{3}} : b^{\frac{1}{3}}$.

21. Show that of all circular sectors described with a given perimeter, that of maximum area has the arc equal to double the radius.

22. A ship steaming north 12 knots an hour sights another ship 10 miles ahead, steaming east 9 knots. What will be the least distance between the ships if each keeps on her course, and at what time will it occur?

Ans. Time, 32 min.; distance, 6 miles.

23. What sector must be taken from a given circle that it may form the curved surface of a cone of maximum volume?

Ans. $\sqrt{\frac{2}{3}}$ of the circle.

24. A Norman window, consisting of a rectangle surmounted by a semicircle, is to admit the maximum amount of light with a given perimeter. Show that the base of the rectangle must be double its altitude.

CHAPTER X.

INDETERMINATE FORMS.

70. Let us consider the fraction

$$\phi(x) = \frac{x^2 - 9}{x - 3}. \qquad (1)$$

For any value we may assign to x there will be a definite value of $\phi(x)$ found by dividing the numerator of the fraction by the denominator.

To this statement there is one exception, the case of $x = 3$. Assigning this value to x, we have

$$\phi(3) = \tfrac{0}{0}.$$

Now, the quotient of two zeros is essentially indeterminate. For the quotient of any two quantities is that quantity which, multiplied by the divisor, will produce the dividend. But any quantity whatever when multiplied by 0 will produce 0. Hence, when divisor and dividend are both zero, any quantity whatever may be their quotient.

But when we consider the terms of the fraction, not as absolute zeros, but as quantities approaching zero as a limit, then their quotient may approach a definite limit. We then regard this limit as the value of the fraction corresponding to zero values of its terms.

As another example, consider the quantity

$$F(x) = \frac{1}{x - 2} - \frac{4}{x^2 - 4}. \qquad (2)$$

We may compute the value of this expression for any value of x except 2. When $x = 2$ the terms will both become infinite. Since if any quantity whatever be added to an infinite

INDETERMINATE FORMS. 129

the sum will be infinite, it follows that any quantity whatever may be the difference of two infinites.

There are several other indeterminate forms. The following are the principal ones which take an algebraic form:

$$\frac{0}{0}; \quad \frac{\infty}{\infty}; \quad 0 \times \infty; \quad \infty - \infty; \quad 0^0; \quad \infty^0; \quad 1^\infty.$$

71. *Evaluation of the Form $\frac{0}{0}$.* In many cases the indeterminate character of an expression may be removed by algebraic transformation. For example, dividing both terms of the fraction (1) by $x - 3$, it becomes $x + 3$, a determinate quantity even for $x = 3$. Again, the expression (2) can be reduced to the form $\frac{1}{x+2}$, which becomes $\frac{1}{4}$ when $x = 2$.

The general method of dealing with the first form is as follows: Let the given fraction be

$$\frac{\phi(x)}{\psi(x)},$$

and let it be supposed that both terms of this fraction vanish when $x = a$, so that we have

$$\phi(a) = 0 \quad \text{and} \quad \psi(a) = 0. \tag{3}$$

Put $h \equiv x - a$, and develop the terms in powers of h by Taylor's theorem. We shall then have

$$\phi(x) = \phi(a + h) = \phi(a) + h\phi'(a) + \frac{h^2}{1 \cdot 2}\phi''(a) + \cdots;$$

$$\psi(x) = \psi(a + h) = \psi(a) + h\psi'(a) + \frac{h^2}{1 \cdot 2}\psi''(a) + \cdots;$$

whence, for the value of the fraction (comp. Eq. (3)),

$$\frac{\phi(x)}{\psi(x)} = \frac{\phi'(a) + \frac{h}{1 \cdot 2}\phi''(a) + \cdots}{\psi'(a) + \frac{h}{1 \cdot 2}\psi''(a) + \cdots}. \tag{4}$$

Now, when h approaches zero as a limit, the value of this fraction approaches

$$\frac{\phi'(a)}{\psi'(a)}$$

as a limit, which is therefore the required limit of the fraction when both its members approach the limit zero.

It may happen that $\phi'(a)$ and $\psi'(a)$ both vanish. In this case the required limit of the fraction in (4) is seen to be

$$\frac{\phi''(a)}{\psi''(a)}.$$

In general: *The required limit is the ratio of the first pair of derivatives of like order which do not both vanish.*

If the first derivative which vanishes is not of the same order in the two terms,—for example, if, of the two quantities $\phi'(a)$ and $\psi'(a)$, one vanishes and the other does not,—then the limit of the fraction will be zero or infinity according as the vanishing derivative is that of the numerator or denominator.

REMARK. It often happens that the terms of the fraction can be developed in the form (4) without forming the successive derivatives. It will then be simpler to use this development instead of forming the derivatives.

EXAMPLES AND EXERCISES.

1. $\dfrac{x^2 - a^2}{x - a}$ for $x = a$.*

$\phi(x) = x^2 - a^2;\quad \phi'(x) = 2x;\quad \therefore \phi'(a) = 2a;$
$\psi(x) = x - a;\quad \psi'(x) = 1;\quad \therefore \psi'(a) = 1.$

$\therefore \operatorname{lim.}\dfrac{x^2 - a^2}{x - a}(x = a) = 2a,$

a result readily obtained by reducing the fraction to its lowest terms.

2. $\dfrac{\log x}{x - 1}$ for $x = 1$. \qquad Ans. 1.

3. $\dfrac{e^x - e^{-x}}{x}$ for $x = 0$. \qquad Ans. 2.

---

* Using strictly the notation of limits, we should define the quantity sought as the *limit* of the fraction when x *approaches the limit* a. But no confusion need arise from regarding the limit of the fraction as its value for $x = a$, as is customary.

INDETERMINATE FORMS. 131

4. $\dfrac{x - \sin x}{x^3}$ for $(x = 0)$.   Ans. $\tfrac{1}{6}$.

Here the successive derivatives of the terms are:

$\phi'(x) = 1 - \cos x;\quad \phi''(x) = \sin x;\quad \phi'''(x) = \cos x.$
$\psi'(x) = 3x^2;\qquad\quad \psi''(x) = 6x;\quad\ \psi'''(x) = 6.$

The third derivatives are the first ones which do not vanish for $x = 0$.

5. $\dfrac{a^x - b^x}{x}$ for $x = 0$.   Ans. $\log a - \log b = \log \dfrac{a}{b}$.

6. $\dfrac{\tan x - \sin x}{x - \sin x}$ for $x = 0$.   Ans. 3.

7. $\dfrac{x^2}{1 - \cos nx}$ for $x = 0$.   Ans. $\dfrac{2}{n^2}$.

8. $\dfrac{a^x - a}{x - 1}$ for $x = 1$.   Ans. $a \log a$.

9. $\dfrac{a^x - b^x}{x - 1}$ for $x = 1$.   Ans. $a \log a - b \log b$.

10. $\dfrac{\sin x - \sin a}{x - a}$ for $x = a$.   Ans. $\cos a$.

11. $\dfrac{\tan y - \tan a}{\cos^2 y - \cos^2 a}$ for $y = a$.   Ans. $\dfrac{\sec^2 a}{2 \sin a}$.

12. $\dfrac{\log(1+x) + \log(1-x)}{\cos x - \sec x}$ for $x = 0$.   Ans. $+1$.

13. $\dfrac{\log(a+x) - \log(a-x)}{x}$ for $x = 0$.   Ans. $\dfrac{2}{a}$.

14. $\dfrac{\sin 2x + 2 \sin^2 x - 2 \sin x}{\cos x - \cos^2 x}$ for $x = 0$.   Ans. 4.

15. $\dfrac{e^x - e^{-x} - 2x}{x - \sin x}$ for $x = 0$.   Ans.

16. $\dfrac{e^y + \sin y - 1}{\log(1+y)}$ for $y = 0$.   Ans. 2.

17. $\dfrac{1 - \sin x - \cos x + \log(1+x)}{e^x - 1 - x}$ for $(x = 0)$.   Ans. 0.

72. Forms $\frac{\infty}{\infty}$ and $0 \times \infty$. These forms may be reduced to the preceding one by a simple transformation. Any fraction $\frac{N}{D}$ may be written in the form $\frac{1 \div D}{1 \div N}$. If N and D both become infinite, $1 \div D$ and $1 \div N$ will both become infinitesimal, and thus the indeterminate form of the fraction will be $\frac{0}{0}$.

Again, if of two factors A and B, A becomes infinitesimal while B becomes infinite, we write the product in the form $\frac{A}{1 \div B}$, and then it is a fraction of the first form.

But this transformation cannot always be successfully applied unless the term which becomes infinite does so through having a denominator which vanishes. For example, let it be required to find the limit of

$$x^m (\log x)^n$$

for $x \doteq 0$. Here x^m approaches zero, while $\log x$, and therefore $(\log x)^n$, becomes infinite for $x = 0$. Hence the denominator of the transformed fraction will be $\frac{1}{l^n}$ (putting for brevity $l \equiv \log x$). The successive derivatives of this quantity with respect to x are

$$\frac{-n}{xl^{n+1}}; \quad \frac{n}{x^2}\left(\frac{1}{l^{n+1}} + \frac{n+1}{l^{n+2}}\right); \quad \text{etc.}$$

The successive derivatives of the numerator are

$$mx^{m-1}; \quad m(m-1)x^{m-2}; \quad \text{etc.}$$

The limiting values of the given quantity $x^m l^n$ thus become

$$-\frac{mx^m l^{n+1}}{n}; \quad \frac{m(m-1)x^m}{n\left(\frac{1}{l^{n+1}} + \frac{n+1}{l^{n+2}}\right)}; \quad \text{etc.,}$$

which remain indeterminate in form how far soever we may carry them.

In such cases the required limit of the fraction can be found only by some device for which no general rule can be laid down. In the example just given the device consists in replacing x by a new variable y, determined by the equation
$$\log x = -y.$$
We then have
$$x = e^{-y}.$$
Since for $x \doteq 0$ $y \doteq \infty$, we now have to find the limit of
$$\frac{(-y)^n}{e^{my}} = (-1)^n \frac{y^n}{e^{my}}$$
for $y \doteq \infty$.

By taking the successive derivatives of the two terms of the fraction $\frac{y^n}{e^{my}}$ we have the successive forms
$$\frac{ny^{n-1}}{me^{my}}; \quad \frac{n(n-1)y^{n-2}}{m^2 e^{my}}; \quad \frac{n(n-1)(n-2)y^{n-3}}{m^3 e^{my}}; \quad \text{etc.}$$

Whatever the value of n, we must ultimately reach an exponent in the numerator which shall be zero or negative, and then the numerator will become $n!$ if n is a positive integer, and will vanish for $y \doteq \infty$, if n is not a positive integer. But the denominator will remain infinite. We therefore conclude:
$$\lim \left[x^m (\log x)^n \right] (x \doteq 0) = 0,$$
whatever be m and n, so long as m is positive.

From this the student should show, by putting $z \equiv x^{-1}$ and $m = 1$, that the fraction
$$\frac{z}{(\log z)^n}$$
becomes infinite with z, how great soever the exponent n, and therefore that *any infinite number is an infinity of higher order than any power of its logarithm.*

73. *Form* $\infty - \infty$. In this case we have an expression of the form
$$F(x) = u - v,$$

in which both u and v become infinite for some value of x. Placing it in the form
$$F(x) = u\left(1 - \frac{v}{u}\right),$$
we see that $F(x)$ will become infinite with u unless the fraction $\frac{v}{u}$ approaches unity as its limit. When this is the case the expression takes the form $\infty \times 0$ of the preceding article.

74. *Form* 1^∞. To investigate this form let us find the limit of the expression
$$\left(1 + \frac{1}{n}\right)^{hn} \equiv u$$
when n becomes infinite. Taking the logarithm, we have
$$\log u = hn \log\left(1 + \frac{1}{n}\right)$$
$$= hn \left\{\frac{1}{n} - \frac{1}{2n^2} + \frac{1}{3n^3} - \cdots\right\}$$
$$= h\left(1 - \frac{1}{2n} + \cdots\right).$$
Making n infinite, we have
$$\lim. \log u = h;$$
or, because the limit of $\log u$ is the logarithm of lim. u,
$$\log \lim. u = h.$$
Hence
$$\lim. \left(1 + \frac{1}{n}\right)^{hn} (n \doteq \infty) = e^h.$$

In order that this result may be finite, h itself must not be infinite. We therefore reach the general conclusion:

THEOREM. *In order that an expression of the form*
$$(1 + \alpha)^x$$
may have a finite limit when α becomes infinitesimal and x infinite, the product αx must not become infinite.

Cor. If the product αx approaches zero as a limit, the given expression will approach the limit unity.

INDETERMINATE FORMS.

75. *Forms 0^0 and ∞^0.* Let an expression taking either of these forms as a limit be represented by $u^\phi \equiv F$. The problem is to find the limiting value of the expression when ϕ approaches zero and u either approaches zero or becomes infinite.

From the identity $\quad u = e^{\log u}$
we derive $\quad F = u^\phi = e^{\phi \log u}$.

We infer that the limit of F will depend upon that of $\phi \log u$.

If lim. $\phi \log u$ is $+\infty$, then lim. $F = \infty$.
If lim. $\phi \log u$ is $-\infty$, then lim. $F = 0$.
If lim. $\phi \log u$ is 0, then lim. $F = 1$.
If lim. $\phi \log u$ is finite, then lim. F is finite.

Hence the rule: *To find the limit of u^ϕ when $\phi \doteq 0$ and $u \doteq 0$ or ∞, put $l \equiv$ lim. $\phi \log u$.* Then

$$\text{lim. } u^\phi = e^l.$$

EXAMPLES AND EXERCISES.

1. Find lim. x^x for $x \doteq 0$.
   Here $\qquad x^x = e^{x \log x}$.

Since $x \log x$ has zero as its limit when $x = 0$, the required limit is e^0 or 1.

2. lim. x^{nx} for $x \doteq 0$. $\qquad$ Ans. $F = 1$.
3. lim. $x^{\frac{1}{x}}$ for $x \doteq \infty$. $\qquad$ Ans. $F = 1$.
4. $x^{\frac{1}{1-x}}$ for $x = 1$. $\qquad$ Ans. $\dfrac{1}{e}$.
5. $x^{\frac{n}{1-x}}$ for $x = 1$. $\qquad$ Ans. e^{-n}.
6. $(1-x)^{\frac{h}{x}}$ for $x = 0$. $\qquad$ Ans. e^{-h}.
7. $\dfrac{e^x - e^{-x}}{\log(1+x)}$ for $x = 0$. $\qquad$ Ans. 2.
8. $\dfrac{\log \sin 2x}{\log \sin x}$ for $x = 0$. $\qquad$ Ans. $= 1$
9. $\dfrac{e^x + \log(1-x) - 1}{x - \tan x}$ for $x = 0$. $\quad$ Ans. $\frac{1}{2}$.

10. $\left(\dfrac{\log x}{x}\right)^{\frac{1}{x}}$ for $x = \infty$.   Ans. 1.

11. $x \tan x - \dfrac{\pi}{2} \sec x$ for $x = \dfrac{\pi}{2}$.   Ans. -1.

12. $y \sin \dfrac{a}{y}$ for $y = \infty$.   Ans. a.

13. $x\left(a^{\frac{1}{x}} - 1\right)$ for $x = \infty$.   Ans. $\log a$.

14. $\left(\dfrac{\tan x}{x}\right)^{\frac{1}{x}}$ for $x = 0$.   Ans. 1.

15. $\left(\dfrac{\tan x}{x}\right)^{\frac{1}{x^2}}$ for $x = 0$.   Ans. $e^{\frac{1}{3}}$.

16. $(\cos x)^{\frac{1}{x^2}}$ for $x = 0$.   Ans. $e^{-\frac{1}{2}}$.

17. $(1-y) \tan \dfrac{\pi}{2} y$ for $y = 1$.   Ans. $\dfrac{2}{\pi}$.

18. $\left(\dfrac{\log x}{x}\right)^{\frac{1}{x}}$ for $x = 0$.   Ans. 1.

19. $x - x^2 \log\left(1 + \dfrac{1}{x}\right)$ for $x = \infty$.   Ans. $\frac{1}{2}$.

20. $\dfrac{e^x - e^{-x}}{\log(1+x)}$ for $x = 0$.   Ans. 2.

21. $\left(\dfrac{a_1^x + a_2^x}{2}\right)^{\frac{2}{x}}$ for $x = 0$.   Ans. $a_1 a_2$.

22. $\left(\dfrac{a_1^x + a_2^x + \ldots + a_n^x}{n}\right)^{\frac{n}{x}}$ for $x = 0$.   Ans. $a_1 a_2 \ldots a_n$.

23. Show that, how great soever the exponent n,
$$\dfrac{x}{(\log x)^n} \doteq \infty \text{ when } x \doteq \infty.$$

CHAPTER XI.

OF PLANE CURVES.

76. *Forms of the Equations of Curves.* As we have heretofore considered curve lines, they have been defined by an equation between the co-ordinates of each point of the curve, and therefore of one of the forms

$$y = f(x); \quad x = f(y); \tag{1}$$

and $$F(x, y) = 0.$$

The distinguishing feature of the equation is that when we assign a value at pleasure to one of the co-ordinates x or y, one or more *corresponding* values of the other co-ordinate are determined by the equation.

But the relation between x and y may be equally well defined by expressing each of them as a function of an auxiliary variable, which is then the independent variable. Calling this auxiliary variable u, the equations of a curve will be of the form

$$\left. \begin{array}{l} x = \phi_1(u); \\ y = \phi_2(u). \end{array} \right\} \tag{2}$$

Assigning values at pleasure to u, we shall have corresponding values of x and y determining each point of the curve.

An advantage of this method of representation is that for each value of u we have one definite point of the curve, or several definite points when the equations give several values of the co-ordinates for each value of u; and we thus have a relation between a point and the algebraic quantity u.

It is also to be remarked that by eliminating u from the equations (2) we shall get a single equation between x and y which will be the equation of the curve in one of the forms (1).

EXAMPLE 1. Let us put
$a, b \equiv$ the co-ordinates of any fixed point B of a straight line;
$\alpha \equiv$ the angle which the line makes with the axis of x;
$\rho \equiv$ the distance of any point P of the line from the point (a, b).

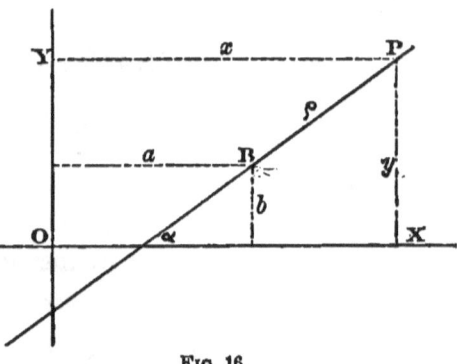

Fig. 16.

Then we readily see from the figure that the co-ordinates x and y of P are given by the equations

$$\left.\begin{array}{l} x = a + \rho \cos \alpha; \\ y = b + \rho \sin \alpha; \end{array}\right\} \qquad (3)$$

which are equations of the straight line in the independent form.

Here ρ is the auxiliary variable, called u in Eq. 2. By eliminating this quantity we shall have

$$x \sin \alpha - y \cos \alpha = a \sin \alpha - b \cos \alpha,$$

which is the equation of the line in one of its usual forms.

EXAMPLE 2. The equation of a circle may be expressed in the form

$$\left.\begin{array}{l} x = a + c \cos u; \\ y = b + c \sin u; \end{array}\right\} \qquad (4)$$

u being the independent variable.

By writing (4) in the form

$$x - a = c \cos u,$$
$$y - b = c \sin u,$$

and eliminating u by taking the sum of the squares of the two equations, we have

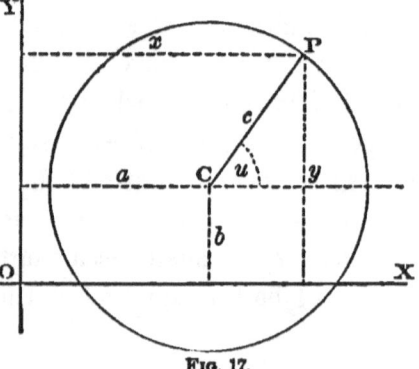

Fig. 17.

$$(x-a)^2 + (y-b)^2 = c^2,$$

the equation of a circle of radius c.

Notice the beautiful relation between (3) and (4). They are the same in form: if in (4) we write ρ for c and α for u, they will be the same equations. Then, by supposing ρ constant and α variable, we are carried round the point (a, b) at a constant distance ρ, that is, around a circle. By supposing ρ variable and α constant we are carried through (a, b) in a constant direction, that is, along a straight line.

77. *Infinitesimal Elements of Curves.* Let P and P' be two points on a curve, P being supposed fixed, and P' variable. We may then suppose P' to approach P as its limit, and inquire into the limits of any magnitudes associated with the curve.

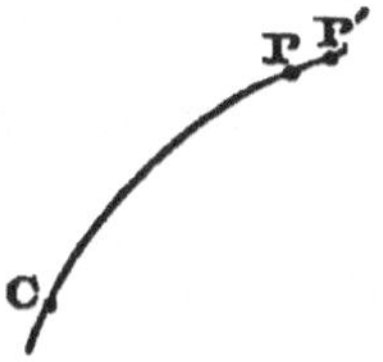

We may also measure the length of an arc of the curve from an initial point C to a terminal point P. Then, supposing C fixed and P variable, PP' may be taken as an increment of the arc.

Fig. 18.

If we put
$$s \equiv \text{arc } CP,$$
we shall have
$$\Delta s = \text{arc } PP'.$$

AXIOM. *The ratio of an infinitesimal element of a curve to the straight line joining its extremities approaches unity as its limit.*

We call this proposition an axiom because a really rigorous demonstration does not seem possible. Its truth will appear by considering that if the curve has no sharp turns, which we presuppose, then it can change its direction only by an infinitesimal quantity in any infinitesimal portion of its length. Now, a line which has the same direction throughout its length is a straight line.

78. Theorem I. *If a straight line touch a curve at the point P, a point P'' on the curve at an infinitesimal distance will, in general, be distant from the tangent by an infinitesimal of the second order.*

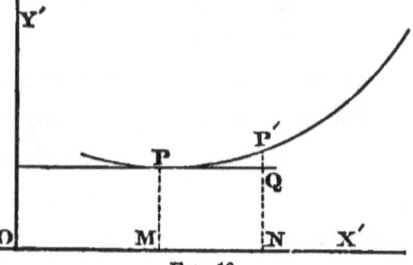

Fig. 19.

Let $y = f(x)$ be the equation of the curve.

Let us transform the equation to a new system of co-ordinates, x' and y', so taken that the axis of X' shall be parallel to the tangent at P. This will make $\dfrac{dy'}{dx'} = 0$. Let x' and y' be the co-ordinates of P, and $(x' + h, y'')$ the co-ordinates of a point P'' near P.

Developing by Taylor's theorem, we have

$$y'' - y' = \frac{dy'}{dx'}h + \frac{d^2y'}{dx'^2}\frac{h^2}{1\cdot 2} + \cdots$$

Now, $y'' - y'$ is the distance $P''Q$ of the point P' from the tangent at P. Since $\dfrac{dy'}{dx'} = 0$, when h becomes infinitesimal the term of highest order in this distance is $\dfrac{d^2y'}{dx'^2}\dfrac{h^2}{1\cdot 2}$, a quantity of the second order.

Remark. In the special case when $\dfrac{d^2y'}{dx'^2} = 0$, the distance in question may be a quantity of the third or of some higher order, according to the order of the first differential coefficient which does not vanish.

Corollary. *The cosine of an infinitesimal arc differs from unity by an infinitesimal of the second order.*

For if we draw a unit circle with its tangent at the initial point, the cosine of an arc will differ from unity by the distance from the end of the arc to the tangent line. When the arc is infinitesimal, the corollary follows from the theorem.

PLANE CURVES.

Theorem II. *The area included between an infinitesimal arc and its chord is not greater than an infinitesimal of the third order.*

From Th. I. we may readily see that the maximum distance between the chord and its arc is a quantity of the second order. The area is less than the product of this distance by the length of the chord, which product is an infinitesimal of at least the third order.

79. *Expressions for Elements of Curves. Def.* **An element** of a geometric magnitude is an infinitesimal portion of that magnitude.

The word implies that we conceive the magnitude to be made up of infinitesimal parts.

Element of an Arc. Let us put

$s \equiv$ the length of any arc of a curve;
$ds \equiv$ an element of this arc.

If P and P' be two points of a curve, we shall have

(chord PP')$^2 = \Delta x^2 + \Delta y^2$.

When PP' becomes infinitesimal, the ratio of ds to PP' becomes unity (§ 77), and we have

$ds^2 = dx^2 + dy^2$;

$ds = \sqrt{dx^2 + dy^2} = \sqrt{1 + \left(\dfrac{dy}{dx}\right)^2}\, dx.$

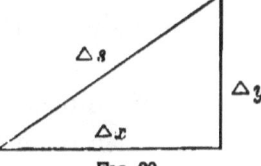

Fig. 20.

Case of Polar Co-ordinates. To express the element of a curve referred to polar co-ordinates, differentiate the equations

$x = r \cos \theta; \quad y = r \sin \theta.$

Thus $dx = \cos \theta\, dr - r \sin \theta\, d\theta;$
$dy = \sin \theta\, dr + r \cos \theta\, d\theta;$

which gives $ds^2 = dr^2 + r^2 d\theta^2$

and $ds = \sqrt{r^2 + \left(\dfrac{dr}{d\theta}\right)^2}\, d\theta.$

80. *Equations of certain Noteworthy Curves. The Cycloid.* The cycloid is a curve described by a point on the circumference of a circle rolling on a straight line. A point on the circumference of a carriage-wheel, as the carriage moves, describes a series of cycloids, one for each revolution of the wheel.

To find the equation of the cycloid, let P be the generating point. Let us take the line on which the circle rolls as the axis of X, and let us place the origin at the point O where P is in contact with the line OX.

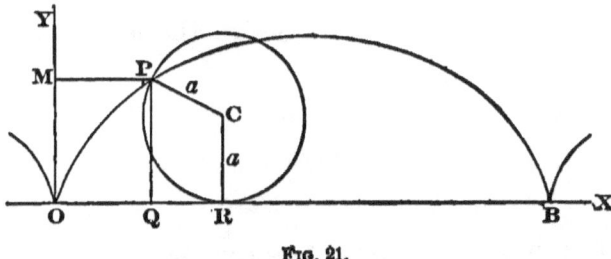

Fig. 21.

Also put

$a \equiv$ the radius of the circle;

$u \equiv$ the angle through which the circle has rolled, expressed in terms of unit-radius.

Then, when the circle has rolled through any distance OR, this distance will be equal to the length of the arc PR of the circle between P and the point of contact R, that is, to au. We thus have, for the co-ordinates of the centre, C, of the circle,

$$x = au;$$
$$y = a;$$

and for the co-ordinates of the point P on the cycloid,

$$\left. \begin{array}{l} x = au - a \sin u = a(u - \sin u); \\ y = a - a \cos u = a(1 - \cos u); \end{array} \right\} \quad (1)$$

which are the equations of the cycloid with u as an independent variable.

PLANE CURVES. 143

To eliminate u, find its value from the second equation,

$$u = \cos^{(-1)}\left(1 - \frac{y}{a}\right).$$

This gives $\sin u = \sqrt{1 - \cos^2 u} = \dfrac{\sqrt{2ay - y^2}}{a}.$

Then, by substituting in the first equation

$$x = a \cos^{(-1)} \frac{a - y}{a} - \sqrt{2ay - y^2}, \qquad (2)$$

which is the equation of the cycloid in the usual form.

81. *The Lemniscate* is the locus of a point, the product of whose distances from two fixed points (called *foci*) is equal to the square of half the distance between the foci.

Let us take the line joining the foci as the axis of X, and the middle point of the segment between the foci as the origin. Let us also put $c \equiv$ half the distance between the foci.

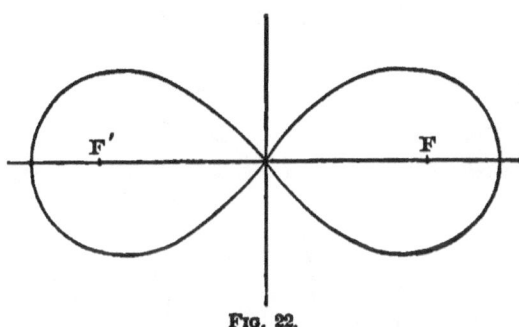

Fig. 22.

Then the distances of any point (x, y) of the curve from the foci are

$$\sqrt{(x - c)^2 + y^2} \quad \text{and} \quad \sqrt{(x + c)^2 + y^2}.$$

Equating the product of these distances to c^2, squaring and reducing, we find

$$(x^2 + y^2)^2 = 2c^2(x^2 - y^2), \qquad (3)$$

which is the equation of the lemniscate.

Transforming to polar co-ordinates by the substitutions
$$x = r\cos\theta,$$
$$y = r\sin\theta,$$
we find, for the polar equation of the lemniscate,
$$r^2 = 2c^2 \cos 2\theta. \tag{4}$$
Putting $y = 0$, we find, for the point in which the curve cuts the line joining the foci,
$$x = \pm(\sqrt{2})c \equiv a.$$
The line a is the semi-axis of the lemniscate. Substituting it instead of c, the rectangular and polar equations of the curve will become
$$\left.\begin{array}{r}(x^2 + y^2)^2 = a^2(x^2 - y^2); \\ r^2 = a^2 \cos 2\theta. \end{array}\right\} \tag{5}$$

82. *The Archimedean Spiral.* This curve is generated by the uniform motion of a point along a line revolving uniformly about a fixed point.

To find its polar equation, let us take the fixed point as the pole, and the position of the revolving line when the generating point leaves the pole as the axis of reference. Let us also put

$a \equiv$ the distance by which the generating point moves along the radius vector while the latter is turning through the unit radius.

Then, when the radius vector has turned through the angle θ, the point will have moved

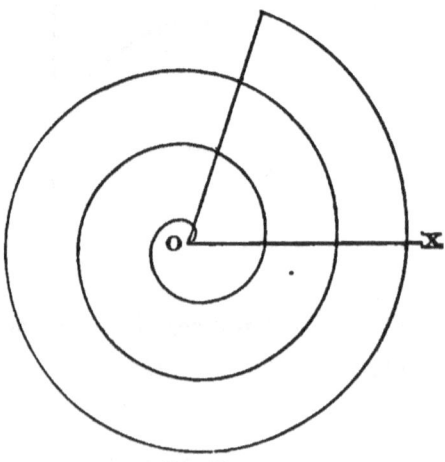

Fig. 23.

from the pole through the distance $a\theta$. Hence we shall have
$$r = a\theta$$
as the polar equation of the Archimedean spiral.

PLANE CURVES. 145

If we increase θ by an entire revolution (2π), the corresponding increment of r will be $2\pi a$, a constant. Hence:

The Archimedean spiral cuts any fixed position of the radius vector in an indefinite series of equidistant points.

83. *The Logarithmic Spiral.* This is a spiral in which the logarithm of the radius vector is proportional to the angle through which the radius vector has moved from an initial position. Hence, if we put θ_0 for the initial angle, we have

$$\log r = l(\theta - \theta_0),$$

l being a constant. Hence

$$r = e^{l\theta - l\theta_0} = e^{-l\theta_0}e^{l\theta}.$$

Putting, for brevity,

$$a \equiv e^{-l\theta_0},$$

the equation of the logarithmic spiral becomes

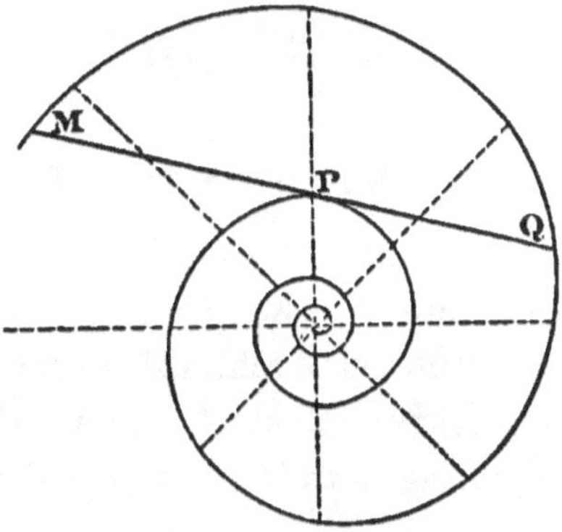

$$r = ae^{l\theta},$$

Fig. 24.

a and l being constants.

EXERCISES.

1. Show (1) that the maximum ordinate of the lemniscate is $\frac{1}{2}c$, and (2) that the circle whose diameter is the line joining the foci cuts the lemniscate at the points whose ordinates are a maximum.

2. Find the following expression for the square of the distance of a point of a cycloid from the starting point (O, Fig. 21):

$$r^2 = 2ay + 2uax - a^2u^2.$$

3. A wheel makes one revolution a second around a fixed axis, and an insect on one of the spokes crawls from the centre toward the circumference at the rate of one inch a second. Find the equation of the spiral along which he is carried.

4. If, in that logarithmic spiral for which $a = 1$ and $l = 1$,
$$r = e^\theta,$$
the radius vector turns through an arc equal to log 2, its length will be doubled.

5. If, in any logarithmic spiral, one radius vector bisects the angle between two others, show that it is a mean proportional between them.

6. Show that the pair of equations
$$x = au^2,$$
$$y = bu,$$
represent a parabola whose parameter is $\dfrac{b^2}{a}$.

7. If, in the equation of the Archimedean spiral, θ and therefore r take all negative values, show that we shall have another Archimedean spiral intersecting the spiral given by positive values of θ in a series of points lying on a line at right angles to the initial position of the revolving line.

This should be done in two ways. Firstly, by drawing the continuation of the spiral when, by a negative rotation of the revolving line, the generating point passes through the pole. It will then be seen that the combination of the two spirals is symmetrical with respect to the vertical axis. Secondly, by expressing the rectangular co-ordinates of a point of the spiral in terms of θ we have
$$x = a\theta \cos \theta,$$
$$y = a\theta \sin \theta.$$
Changing the sign of θ in this equation will change the sign of x and leave y unchanged.

8. Show that if we draw two lines through the centre of a lemniscate making angles of 45° with the axes, no point of the curve will be contained between these lines and the axis of Y.

CHAPTER XII.

TANGENTS AND NORMALS.

84. A **tangent** to a curve is a straight line through two coincident points of the curve.

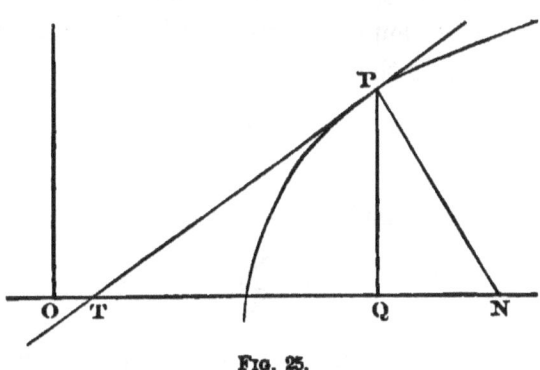

Fig. 25.

A **normal** is a straight line through a point of the curve perpendicular to the tangent at that point.

The **subtangent** is the projection, TQ, upon the axis of X, of that segment TP of the tangent contained between the point of contact and the axis of X.

The **subnormal** is the corresponding projection, QN, of the segment PN of the normal.

Notice that a tangent and a normal are lines of indefinite length, while the subtangent and subnormal are segments of the axis of abscissas. Hence the former are determined by their equations, which will be of the first degree in x and y, while the latter are determined by algebraic expressions for their length.

But the segments TP and PN are sometimes taken as lengths of the tangent and normal respectively, when we consider these lines as segments.

85. General Equation for a Tangent. The general problem of tangents to a curve may be stated thus:

To find the condition which the parameters of a straight line must satisfy in order that the line may be tangent to a given curve.

But it is commonly considered in the more restricted form:
To find the equation of a tangent to a curve at a given point on the curve.

Let (x_1, y_1) be the given point on the curve. By Analytic Geometry the equation of any straight line through this point may be expressed in the form

$$y - y_1 = m(x - x_1); \tag{5}$$

m being the tangent of the angle which the line makes with the axis of X. But we have shown (§ 20) that

$$m = \frac{dy_1}{dx_1},$$

this differential coefficient being formed by differentiating the equation of the curve. Hence

$$y - y_1 = \frac{dy_1}{dx_1}(x - x_1) \tag{6}$$

is the equation of the tangent to any curve at a point (x_1, y_1) on the curve.

Equation of the Normal. The normal at the point (x_1, y_1) passes through this point, and is perpendicular to the tangent. If m' be its slope, the condition that it shall be perpendicular to the tangent is (An. Geom.)

$$m' = -\frac{1}{m} = -\frac{1}{\frac{dy_1}{dx_1}}.$$

Hence the equation of the normal at the point (x_1, y_1) is

$$\frac{dy_1}{dx_1}(y - y_1) = x_1 - x. \tag{7}$$

TANGENTS AND NORMALS. 149

In these equations of the tangent and normal it is necessary to distinguish between the cases in which the symbols x and y represent the co-ordinates of points on the tangent or normal line, and those where they represent the given point of the curve. Where both enter into the same equation, one set, that pertaining to the curve, must be marked by suffixes or accents.

86. *Subtangent and Subnormal.* To find the length of the subtangent and subnormal, we have to find the abscissa x_0 of the point T in which the tangent cuts the axis of abscissas. We then have, by definition,

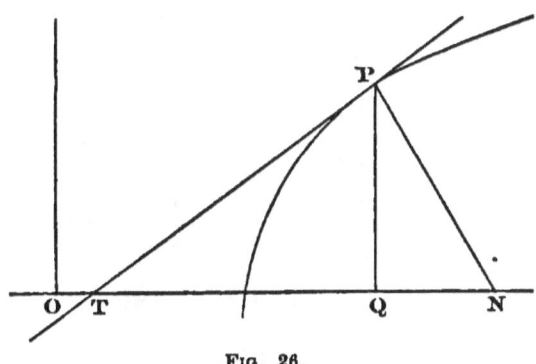

Fig. 26.

$$\text{Subtangent} = x_1 - x_0$$

The value of x_0 is found by putting $y = 0$ and $x = x_0$ in the equation of the tangent. Thus, (6) gives

$$- y_1 = \frac{dy_1}{dx_1}(x_0 - x_1).$$

Hence, for the length of the subtangent TQ,

$$\text{Subtangent} = x_1 - x_0 = \frac{y_1}{\frac{dy_1}{dx_1}}. \tag{8}$$

We find in the same way from (7), for QN,

$$\text{Subnormal} = - y_1\frac{dy_1}{dx_1}. \tag{9}$$

87. Modified Forms of the Equation. In the preceding discussion it is assumed that the equation of the curve is given in the form

$$y = f(x).$$

But, firstly, it may be given in the form

$$F(x, y) = 0.$$

We shall then have (§ 37)

$$\frac{dy_1}{dx_1} = -\frac{\dfrac{dF}{dx_1}}{\dfrac{dF}{dy_1}}.$$

Substituting this value in the equations (6) and (7), we find

$$\left.\begin{array}{l} \text{Tangent: } \dfrac{dF}{dy_1}(y - y_1) = \dfrac{dF}{dx_1}(x_1 - x); \\[2mm] \text{Normal: } \dfrac{dF}{dx_1}(y - y_1) = \dfrac{dF}{dy_1}(x - x_1). \end{array}\right\} \quad (10)$$

Secondly, if the curve is defined by two equations of the form

$$\left.\begin{array}{l} x = \phi_1(u), \\ y = \phi_2(u), \end{array}\right\} \quad (11)$$

we have

$$\frac{dy_1}{dx_1} = \frac{\dfrac{dy}{du}}{\dfrac{dx}{du}},$$

in which there is no need of suffixes to x and y in the second member, because this member is a function of u, which does not contain x or y.

By substitution in (6) and (7), we find

$$\left.\begin{array}{l} \text{Eq. of tangent: } (y - y_1)\dfrac{dx}{du} = (x - x_1)\dfrac{dy}{du}. \\[2mm] \text{Eq. of normal: } (y - y_1)\dfrac{dy}{du} = (x_1 - x)\dfrac{dx}{du}. \end{array}\right\} \quad (12)$$

By substituting in these equations for x_1, y_1, $\dfrac{ax}{dy}$ and $\dfrac{dy}{du}$ their values in terms of u, the parameters of the lines will be functions of u. Then, for each value we assign to u, (11) will give the co-ordinates of a point on the curve, and (12) will determine the tangent and normal at that point.

88. *Tangents and Normals to the Conic Sections.* Writing the equation of the ellipse in the form

$$a^2 y^2 + b^2 x^2 = a^2 b^2, \qquad (a)$$

we readily find, by differentiation,

$$\frac{dy}{dx} = -\frac{b^2 x}{a^2 y}.$$

Applying the suffix to x and y, to show that they represent co-ordinates of points on the ellipse, substituting in (6) and (7), and noting that x_1 and y_1 satisfy (a), we readily find:

For the tangent: $\dfrac{x_1 x}{a^2} + \dfrac{y_1 y}{b^2} = 1.$

For the normal: $\dfrac{a^2}{x_1} x - \dfrac{b^2}{y_1} y = a^2 - b^2.$

Taking the equation of the hyperbola,

$$-a^2 y^2 + b^2 x^2 = a^2 b^2,$$

we find, in the same way,

For the tangent: $\dfrac{x_1 x}{a^2} - \dfrac{y_1 y}{b^2} = 1.$

For the normal: $\dfrac{a^2}{x_1} x + \dfrac{b^2}{y_1} y = a^2 + b^2.$

Taking the equation of the parabola,

$$y^2 = 2px,$$

we find, by a similar process,

For the tangent: $\quad y_1 y = p(x + x_1).$

For the normal: $\quad y - y_1 = -\dfrac{y_1}{p}(x_1 - x).$

152   THE DIFFERENTIAL CALCULUS.

89. Problem. *To find the length of the perpendicular dropped from the origin upon a tangent or normal.*

It is shown in Analytic Geometry that if the equation of a straight line be reduced to the form

$$Ax + By + C = 0,$$

the perpendicular upon the line from the origin is

$$p = \frac{C}{\sqrt{A^2 + B^2}}.$$

It must be noted that in the above form the symbol C represents the sum of all the terms of the equation of the line which do not contain either x or y.

If we have the equation of the line in the form

$$y - y_1 = m(x - x_1),$$

we write it

$$mx - y - mx_1 + y_1 = 0,$$

and then we have

$$A = m;$$
$$B = -1;$$
$$C = y_1 - mx_1.$$

Thus, the expression for the perpendicular is

$$p = \frac{y_1 - mx_1}{\sqrt{m^2 + 1}}.$$

Substituting for m the values already found for the tangent and normal respectively, we find,

For the perpendicular on the tangent:

$$p = \frac{y_1 - x_1 \frac{dy_1}{dx_1}}{\sqrt{1 + \left(\frac{dy_1}{dx_1}\right)^2}} = \frac{y_1 dx_1 - x_1 dy_1}{ds}. \quad (1)$$

For the perpendicular on the normal:

$$p = \frac{x_1 + y_1 \frac{dy_1}{dx_1}}{\sqrt{1 + \left(\frac{dy_1}{dx_1}\right)^2}} = \frac{x_1 dx_1 + y_1 dy_1}{ds}. \quad (2)$$

90. Tangent and Normal in Polar Co-ordinates.

PROBLEM. To find the angle which the tangent at any point makes with the radius vector of that point.

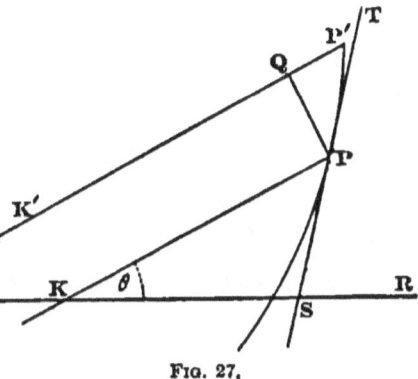

Fig. 27.

Let PP' be a small arc of a curve referred to polar co-ordinates;

KP, a small part of the radius vector of the point P (the pole being too far to the left to be shown in the figure);

$K'P'$, the same for the point P'.

KSR, a parallel to the axis of reference. Drop $PQ \perp K'P'$.

Let SPT be the tangent at P. We also put

$\gamma \equiv$ angle KPS which the tangent makes with the radius vector.

Then let P' approach P as its limit. Then

$$QP' \doteq dr; \quad PQ \doteq rd\theta;$$
$$\tan \gamma \doteq \frac{PQ}{QP'} \doteq \frac{rd\theta}{dr}. \tag{1}$$

We also have

$$\left.\begin{array}{c} \cos \gamma = \dfrac{1}{\sqrt{(1+\tan^2 \gamma)}} = \dfrac{1}{\sqrt{\left\{r^2 + \left(\dfrac{dr}{d\theta}\right)^2\right\}}} \cdot \dfrac{dr}{d\theta}; \\ \\ \sin \gamma = \cos \gamma \tan \gamma = \dfrac{r}{\sqrt{\left\{r^2 + \left(\dfrac{dr}{d\theta}\right)^2\right\}}}. \end{array}\right\} \tag{2}$$

Cor. The angle RSP which the tangent makes with the axis of reference is $\gamma + \theta$.

91. *Perpendicular from the Pole upon the Tangent and Normal.* When γ is the angle between the tangent and the radius vector, we readily find, by geometrical construction, that the perpendicular from the pole upon the tangent and normal are, respectively,

$$p = r \sin \gamma \quad \text{and} \quad p = r \cos \gamma.$$

Substituting for $\sin \gamma$ and $\cos \gamma$ the values already found, we have,

For the perpendicular on tangent:

$$p = \frac{r^2}{\sqrt{\left\{ r^2 + \left(\frac{dr}{d\theta}\right)^2 \right\}}};$$

For the perpendicular on normal:

$$p = \frac{r}{\sqrt{\left\{ r^2 + \left(\frac{dr}{d\theta}\right)^2 \right\}}} \frac{dr}{d\theta}.$$

(3)

92. Problem. *To find the equation of the tangent and normal at a given point of a curve whose equation is expressed in polar co-ordinates.*

It is shown in Analytic Geometry that if we put

$p \equiv$ the perpendicular dropped from the origin upon a line;
$\alpha \equiv$ the angle which this perpendicular makes with the axis of X;

the equation of the line may be written

$$x \cos \alpha + y \sin \alpha - p = 0. \quad (1)$$

Now, as just shown, the tangent makes the angle $\gamma + \theta$ with the axis of X, and the perpendicular dropped upon it makes an angle $90°$ less than this. Hence we have

$$\alpha = \gamma + \theta - 90°;$$
$$\cos \alpha = \sin(\gamma + \theta) = \sin \gamma \cos \theta + \cos \gamma \sin \theta;$$
$$\sin \alpha = -\cos(\gamma + \theta) = -\cos \gamma \cos \theta + \sin \gamma \sin \theta.$$

By substitution in (1), the equation of the tangent becomes
$$x(\sin \gamma \cos \theta + \cos \gamma \sin \theta)$$
$$- y(\cos \gamma \cos \theta - \sin \gamma \sin \theta) - p = 0.$$
Substituting for $\cos \gamma$, $\sin \gamma$ and p the values already found, this equation of the tangent reduces to

$$\left(r \cos \theta + \frac{dr}{d\theta} \sin \theta\right) x + \left(r \sin \theta - \frac{dr}{d\theta} \cos \theta\right) y - r^2 = 0, \quad (2)$$

r and θ being the co-ordinates of the point of tangency.

In the case of the normal the perpendicular upon it is parallel to the tangent. Therefore, to find the equation of the normal, we must put in (1)
$$\alpha = \gamma + \theta.$$

Substituting this value of α, and proceeding as in the case of the tangent, we find, for the normal,

$$\left(\frac{dr}{d\theta} \cos \theta - r \sin \theta\right) x + \left(r \cos \theta + \frac{dr}{d\theta} \sin \theta\right) y - r\frac{dr}{d\theta} = 0. \quad (3)$$

Generally these equations will be more convenient in use if we divide them throughout by r. Thus we have:

Equation of the tangent:
$$\left(\cos \theta + \frac{1}{r}\frac{dr}{d\theta} \sin \theta\right) x + \left(\sin \theta - \frac{1}{r}\frac{dr}{d\theta} \cos \theta\right) y - r = 0. \quad (4)$$

Equation of the normal:
$$\left(\frac{1}{r}\frac{dr}{d\theta} \cos \theta - \sin \theta\right) x + \left(\frac{1}{r}\frac{dr}{d\theta} \sin \theta + \cos \theta\right) y - \frac{dr}{d\theta} = 0. \quad (5)$$

In using these equations it must be noticed that the coefficients of x and y are functions of r and θ, the polar co-ordinates of the point of tangency. When r, θ and $\frac{dr}{d\theta}$ are given, this point and the tangent through it are completely determined.

EXERCISES.

1. Show that in the case of the Archimedean spiral the general expressions for the perpendiculars from the pole upon the tangent and normal, respectively, are

$$p = \frac{a\theta^2}{\sqrt{(1+\theta^2)}} \quad \text{and} \quad p = \frac{a\theta}{\sqrt{(1+\theta^2)}}.$$

Thence define at what point of the spiral the radius vector makes angles of 45° with the tangent and normal. Find also what limit the perpendicular upon the normal approaches as the folds of the spiral are continued out to infinity.

Show also from § 92 that the tangent is perpendicular to the line of reference at every point for which

$$r \sin \theta - a \cos \theta = 0,$$

and hence that, as the folds of the spiral are traced out to infinity, the ordinates of the points of contact of such a tangent approach $\pm a$ as their limit.

2. Show by Eq. 12 that in the case of the logarithmic spiral the angle which the radius vector makes with the tangent is a constant, given by the equation

$$\tan \gamma = \frac{1}{l}.$$

3. Show from Eq. 12 that if a curve passes through the pole, the tangent at that point coincides with the radius vector, unless $\dfrac{dr}{d\theta} = 0$ at this point. Thence show that in the lemniscate the tangents at the origin each cut the axes at angles of 45°.

4. Show that the double area of the triangle formed by a tangent to an ellipse and its axes is $\dfrac{a^2 b^2}{x_1 y_1}$. Then show that the area is a maximum when $\dfrac{x_1}{a} = \pm \dfrac{y_1}{b}$.

Show also that the area of the triangle formed by a normal and the axes is a maximum for the same point.

CHAPTER XIII.

OF ASYMPTOTES, SINGULAR POINTS AND CURVE-TRACING.

93. *Asymptotes.* An **asymptote** of a curve is the limit which the tangent approaches when the point of contact recedes to infinity.

In order that a curve may have a real asymptote, it must extend to infinity, and the perpendicular from the origin upon the tangent must then approach a finite limit.

For the first condition it suffices to show that to an infinite value of one co-ordinate corresponds a real value, finite or infinite, of the other.

For the second condition it suffices to show that the expression for the perpendicular upon the tangent (§§ 89, 91) approaches a finite limit when one co-ordinate of the point of contact becomes infinite. If, as will generally be most convenient, the equation of the curve is written in the form

$$F(x, y) = 0, \qquad (1)$$

the value (1) of the perpendicular, omitting suffixes, may be reduced to

$$p = \frac{y\frac{dF}{dy} + x\frac{dF}{dx}}{\left\{\left(\frac{dF}{dx}\right)^2 + \left(\frac{dF}{dy}\right)^2\right\}^{\frac{1}{2}}}. \qquad (2)$$

If this expression approaches a real finite limit for an infinite value of x or y, the curve has an asymptote.

If the curve is referred to polar co-ordinates, we use the expression (3), § 91, for p. If this approaches a real finite limit for an infinite value of r, the curve has an asymptote.

The existence of the asymptote being thus established, its equation may generally be found from the form (10), § 87, which we may write thus:

$$\frac{dF}{dx_1}x + \frac{dF}{dy_1}y = x_1\frac{dF}{dx_1} + y_1\frac{dF}{dy_1}, \qquad (3)$$

by supposing x_1 or y_1 to become infinite.

Commonly the coefficients $\dfrac{dF}{dx_1}$ and $\dfrac{dF}{dy_1}$ will themselves become infinite with the co-ordinates. We must then divide the whole equation by such powers of x_1 and y_1 that none of the terms shall become infinite.

94. *Examples of Asymptotes.*

1. $F(x) = x^3 + y^3 - 3axy = 0.\,(a)$

The curve represented by this equation is called the *Folium of Descartes*. The equation (3) gives in this case, applying suffixes,

$(x_1^2 - ay_1)x + (y_1^2 - ax_1)y$
$\quad = x_1^3 + y_1^3 - 2ax_1y_1 = ax_1y_1.$

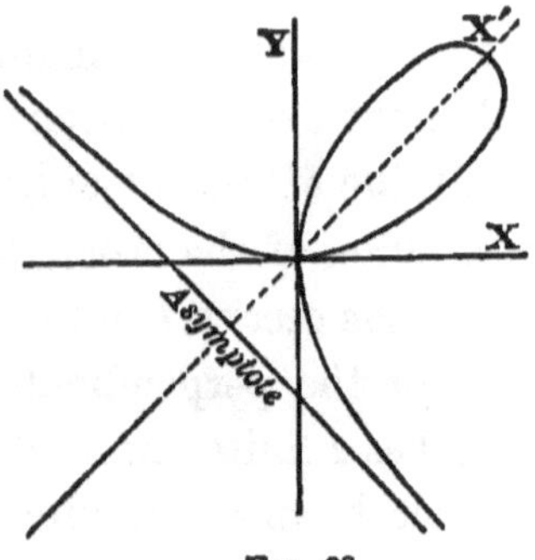

Fig. 28.

To make the coefficients of x and y finite for $x_1 = \infty$, divide by x_1y_1. Then the equation becomes

$$\left(\frac{x_1}{y_1} - \frac{a}{x_1}\right)x + \left(\frac{y_1}{x_1} - \frac{a}{y_1}\right)y - a = 0. \qquad (b)$$

Let us now find from (a) the limit of y_1 for $x_1 \doteq \infty$. We have

$$1 + \frac{y_1^3}{x_1^3} = 3a\frac{y_1}{x_1^2}.$$

The second member of this equation will approach zero as a limit, unless y_1 is an infinite of as high an order as $x_1^{\frac{2}{3}}$, which is impossible, because then the first member of the equation containing y_1^3 would be an infinite of higher order than the second member, which is absurd. Hence, passing to the limit,

$$\lim \left(\frac{y_1}{x_1}\right)(x_1 \doteq \infty) = -1.$$

Then, by substitution in (b), we find, for the asymptote,

$$x + y + a = 0.$$

2. Take next the equation

$$F(x, y) = x^3 - 2x^2y - ax^2 - a^2y = 0. \qquad (a)$$

With this equation (3) becomes

$$(3x_1^2 - 4x_1y_1 - 2ax_1)x - (2x_1^2 + a^2)y$$
$$= 3x_1^3 - 6x_1^2y_1 - 2ax_1^2 - a^2y_1. \qquad (b)$$

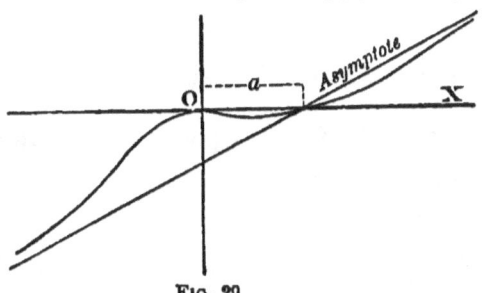

Fig. 29.

We notice that the terms of highest order in the second member are three times those of highest order in (a). From (a) we have

$$x_1^3 - 2x_1^2y_1 = ax_1^2 + a^2y_1.$$

Substituting in the second member of (b), and dividing by x_1^2, (b) becomes

$$\left(3 - 4\frac{y_1}{x_1} - \frac{2a}{x_1}\right)x - \left(2 + \frac{a^2}{x_1^2}\right)y = a + \frac{2a^2y_1}{x_1^2}. \qquad (b')$$

Solving (a) for y, we find

$$\frac{y_1}{x_1} = \frac{x_1^2 - ax_1}{2x_1^2 + a^2},$$

an expression which approaches the limit $\frac{1}{2}$ when $x_1 \doteq \infty$. Thus, passing to the limit, (b') gives, for the equation of the asymptote,

$$x - 2y = a.$$

3. *The Witch of Agnesi*. This curve is named after the Italian lady who first investigated its properties. Its equation is

$$x^2 y + a^2 y - a^3 = 0. \quad (a)$$

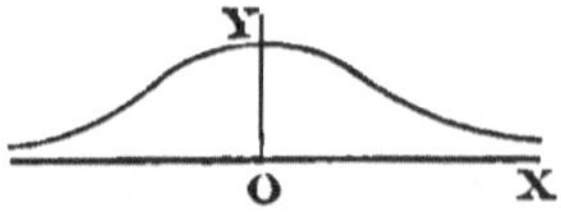

The equation of the tangent is

Fig. 30.

$$2 x_1 y_1 x + (x_1^2 + a^2) y = 3 x_1^2 y_1 + a^2 y_1 = 3 a^3 - 2 a^2 y_1. \quad (b)$$

By solving (a) for x and y respectively we see that x_1 may become infinite, but that y_1 is always positive and less than a. Hence, to make the coefficient of y in (b) finite for $x_1 = \infty$, we must divide by x_1^2, which reduces the equation of the asymptote to

$$y = 0.$$

Hence the axis of x is itself an asymptote.

95. *Points of Inflection.* A point of inflection is a point where the tangent intersects the curve at the point of tangency.

It is evident from the figure that in passing along the curve, and considering the slope of the tangent at each point, the point of inflection is one at which this slope is a maximum or a minimum. Because we have

Fig. 31.

$$\text{slope} = \frac{dy}{dx},$$

the conditions that the slope shall be a maximum or minimum are

$$\frac{d^2 y}{dx^2} = 0$$

and $\dfrac{d^3 y}{dx^3}$ different from zero. If the first condition is fulfilled, but if $\dfrac{d^3 y}{dx^3}$ is also zero, we must proceed, as in problems of maxima and minima, to find the first derivative in order which does not vanish. If the order of this derivative is even, there is no point of inflection for $\dfrac{d^2y}{dx^2} = 0$; if odd, there is one.

As an example, let it be required to find the points of inflection of the curve

$$xy^2 = a^2(a - x).$$

Reducing the equation to the form

$$y^2 = \frac{a^3}{x} - a^2,$$

we find

$$\frac{dy}{dx} = -\frac{a^3}{2x^2y};$$

$$\frac{d^2y}{dx^2} = \frac{a^3}{2x^4y^2}\left(2xy + x^2\frac{dy}{dx}\right) = \frac{a^3}{2x^4y^2}\left(2xy - \frac{a^3}{2y}\right).$$

The condition that this expression shall vanish is

$$4xy^2 = a^3,$$

which, compared with the equation of the curve, gives, for the co-ordinates of the point of inflection,

$$x = \frac{3}{4}a; \quad y = \pm\frac{a}{\sqrt{3}}.$$

EXERCISES.

Find the points of inflection of the following curves:

1. $xy = a^2 \log \dfrac{x}{a}.$ Ans. $\begin{cases} x = ae^{\frac{3}{2}}. \\ y = \frac{3}{2}ae^{-\frac{3}{2}}. \end{cases}$

2. $\begin{cases} x = a(1 - \cos u); \\ y = a(nu + \sin u). \end{cases}$

Ans. $\begin{cases} x = \dfrac{(n+1)a}{n}; \\ y = a\left(\cos^{(-1)}\left(-\dfrac{1}{n}\right) + \dfrac{\sqrt{n^2 - 1}}{n}\right). \end{cases}$

THE DIFFERENTIAL CALCULUS.

96. *Singular Points of Curves.* If we conceive an infinitesimal circle to be drawn round any point of a curve as a centre, then, in general, the curve will cut the circle in two opposite points only, which will be 180° apart.

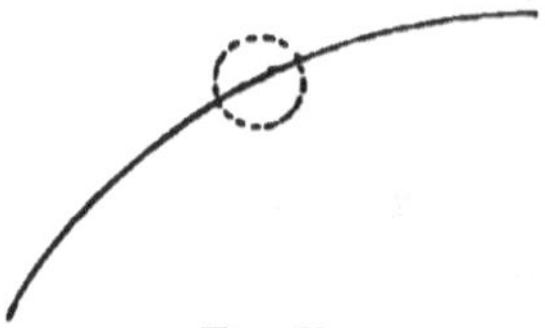

FIG. 32.

But special points may sometimes be found on a curve where the infinitesimal circle will be cut in some other way than this: perhaps in more or less than two points; perhaps in points not 180° apart. These are called **singular points**.

The principal singular points are the following:

Double-points; at which a curve intersects itself. Here the curve cuts the infinitesimal circle in four points (Fig. 33).

Cusps; where two branches of a curve terminate by touching each other (Fig. 34). Here the infinitesimal circle is cut in two coincident points.

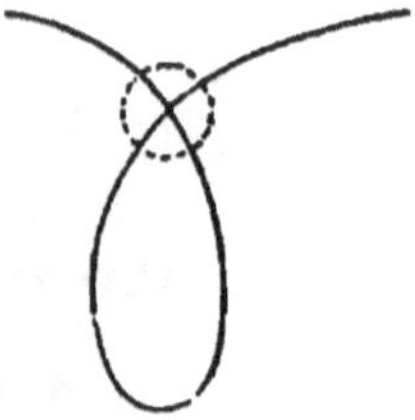

FIG. 33.  FIG. 34.

Stopping Points; where a curve suddenly ends. Here the infinitesimal circle is cut in only a single point.

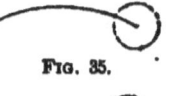
FIG. 35.

Isolated Points; from which no curve proceeds, so that the infinitesimal circle is not cut at all.

FIG. 36.

Salient Points; from which proceed two branches making with each other an angle which is neither zero nor 180°. Here the infinitesimal circle is cut in two points which are neither apposite nor coincident.

There may also be *multiple-points*, through which the curve passes any number of times. A double-point is a special kind of multiple-point.

A multiple-point through which the curve passes three times is called a triple-point.

97. Condition of Singular Points. Let (x_0, y_0) be any point on a curve, and let it be required to investigate the question whether this point is a singular one. We first transform the equation of the curve to one in polar coordinates having the point (x_0, y_0) as the pole. To do this we put, in the equation of the curve,

$$\left. \begin{array}{l} x = x_0 + \rho \cos \theta; \\ y = y_0 + \rho \sin \theta. \end{array} \right\} \quad (1)$$

The resulting equation between ρ and θ will be the

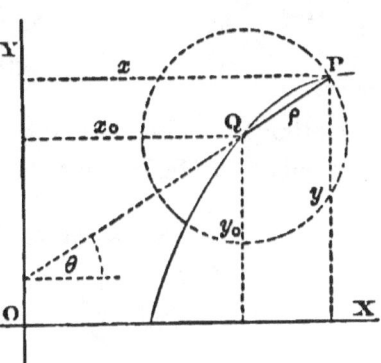

Fig. 37.

equation of the curve referred to (x_0, y_0) as the pole. Moreover, if we assign to ρ a fixed value, the corresponding value of θ derived from the equation will be the angle θ showing the direction QP from Q to the point P, where the circle of radius ρ cuts the curve. The limit which θ approaches as ρ becomes infinitesimal will determine the points of intersection of the infinitesimal circle with the curve.

If, now, the given equation of the curve is

$$F(x, y) = 0,$$

then, by the substitution (1), the polar equation will be

$$F(x_0 + \rho \cos \theta, y_0 + \rho \sin \theta) = 0. \quad (2)$$

Now, let us develop this expression in powers of ρ by Maclaurin's theorem. Since ρ enters into (2) only through x and y in (1), we have

$$\frac{dF}{d\rho} = \frac{dF}{dx}\frac{dx}{d\rho} + \frac{dF}{dy}\frac{dy}{d\rho} = \cos \theta \frac{dF}{dx} + \sin \theta \frac{dF}{dy} \equiv F',$$

$\left(\text{because } \dfrac{dx}{d\rho} = \cos \theta \text{ and } \dfrac{dy}{d\rho} = \sin \theta \right).$

Then

$$\frac{d^2F}{d\rho^2} = \frac{dF'}{d\rho} = \cos\theta\left(\frac{d^2F}{dx^2}\frac{dx}{d\rho} + \frac{d^2F}{dxdy}\frac{dy}{d\rho}\right)$$
$$+ \sin\theta\left(\frac{d^2F}{dxdy}\frac{dx}{d\rho} + \frac{d^2F}{dy^2}\frac{dy}{d\rho}\right)$$
$$= \cos^2\theta\frac{d^2F}{dx^2} + 2\sin\theta\cos\theta\frac{d^2F}{dxdy} + \sin^2\theta\frac{d^2F}{dy^2} \equiv F''$$

Noting that when $\rho = 0$ then $x = x_0$, we see that the development by Maclaurin's theorem will be

$$F(x, y) = F(x_0, y_0) + \rho\left(\cos\theta\frac{dF}{dx_0} + \sin\theta\frac{dF}{dy_0}\right)$$
$$+ \frac{1}{2}\rho^2\left(\cos^2\theta\frac{d^2F}{dx_0^2} + 2\sin\theta\cos\theta\frac{d^2F}{dx_0 dy_0} + \sin^2\theta\frac{d^2F}{dy_0^2}\right)$$
$$+ \text{etc.} \qquad\qquad = 0$$

Here $\frac{dF}{dx_0}$ means the value of $\frac{dF}{dx}$ when x_0 is put for x, etc.

Because (x_0, y_0) is by hypothesis a point on the curve, we have $F(x_0, y_0) = 0$, and the only terms of the second member are those in ρ, ρ^2, etc. Thus the polar equation (2) of the curve may be written

or $$\left.\begin{array}{l} F_0'\rho + F_0''\rho^2 + F_0'''\rho^3 + \text{etc.} = 0, \\ F_0' + F_0''\rho + F_0'''\rho^2 + \text{etc.} = 0. \end{array}\right\} \quad (3$$

To find the points in which the curve cuts a circle of radius ρ, we have to determine θ as a function of ρ from this equation. When ρ is an infinitesimal, all the terms after the first will be infinitesimals. Hence, at the limit, where ρ becomes infinitesimal θ must satisfy the equation

$$F_0' = 0,$$

which gives $$\tan\theta = -\frac{\dfrac{dF}{dx_0}}{\dfrac{dF}{dy_0}}.$$

This is the known equation for the slope of the tangent at (x_0, y_0), and gives only the evident result that in general th

ASYMPTOTES AND SINGULAR POINTS.   165

rve cuts the infinitesimal circle along the line tangent to
e curve at Q.

But, if possible, let the point $(x_0 y_0)$ be so taken that

$$\frac{dF'}{dx_0} = 0; \quad \frac{dF'}{dy_0} = 0. \tag{4}$$

Then we shall have $F_0' = 0$, and the equation (3) of the
rve will reduce to

$$F_0''\rho + F_0'''\rho^2 + \text{etc.} = 0,$$
$$F_0'' + F_0'''\rho + \text{etc.} = 0.$$

Again, letting ρ become infinitesimal, we shall have at the
mit

$$F_0'' = \cos^2\theta \frac{d^2F}{dx_0^2} + 2\sin\theta\cos\theta \frac{d^2F}{dx_0 dy_0} + \sin^2\theta \frac{d^2F}{dy_0^2} = 0. \tag{5}$$

Dividing throughout by $\cos^2\theta$, we shall have a quadratic
quation in $\tan\theta$, which will have two roots. Since each
lue of $\tan\theta$ gives a pair of opposite points in which the
rve may cut the infinitesimal circle, and since (5) depends
n (4), we conclude:

*The necessary condition of a double-point is that the three
quations*

$$F(x, y) = 0, \quad \frac{dF(x, y)}{dx} = 0, \quad \frac{dF(x, y)}{dy} = 0,$$

all be satisfied by a single pair of values of x and y.

If the two values of $\tan\theta$ derived from $F_0'' = 0$ are equal,
e shall have either a cusp, or a point in which two branches
f the curve touch each other. If the roots are imaginary,
e singular point will be an isolated point.

98. *Examples of Double-points.* A curve whose equation
ontains no terms of less than the second degree in x and y
as a singular point at the origin. For example, if the equa-
on be of the form

$$F(x, y) = Px^2 + Qxy + Ry^2 = 0,$$

en this expression and its derivatives with respect to x and
will vanish for $x = 0$ and $y = 0$.

Let us now investigate the double-points of the curve
$$(y^2 - a^2)^2 - 3a^2x^2 - 2ax^3 = 0. \tag{1}$$
We have
$$\left.\begin{aligned} \frac{dF}{dx} &= -6(a^2x + ax^2) = -6ax(a+x); \\ \frac{dF}{dy} &= 4y(y^2 - a^2) = 4y(y+a)(y-a). \end{aligned}\right\} \tag{2}$$

The first of these derivatives vanishes for $x = 0$ or $-a$;
The second of these derivatives vanishes for $y = 0$, $-a$ or $+a$.

Of these values the original equation is satisfied by the following pairs:
$$\left.\begin{aligned} x_0 &= 0; & 0; & -a; \\ y_0 &= -a; & +a; & 0; \end{aligned}\right\} \tag{3}$$
which are therefore the co-ordinates of singular points.

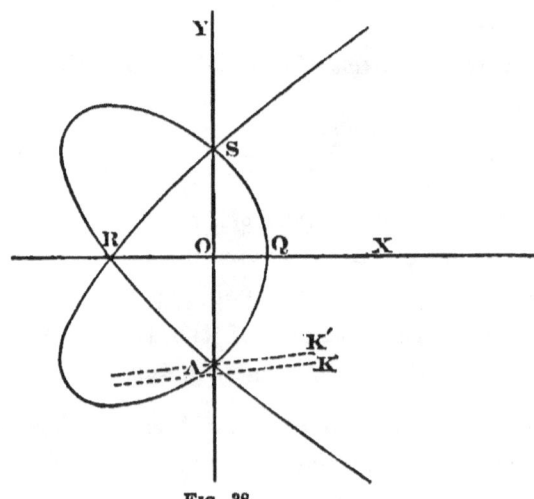

Fig. 38.

Differentiating again, we have
$$\frac{d^2F}{dx^2} = -6a^2 - 12ax; \quad \frac{d^2F}{dxdy} = 0; \quad \frac{d^2F}{dy^2} = 12y^2 - 4a^2.$$

Forming the equation $F'' = 0$, it gives

$$(12y^2 - 4a^2) \tan^2 \theta = 6a^2 + 12ax.$$

Substituting the pairs of co-ordinates (3), we find:

At the point $(0, -a)$, $\tan \theta = \pm \frac{1}{2}\sqrt{3}$;
At the point $(0, +a)$, $\tan \theta = \pm \frac{1}{2}\sqrt{3}$;
At the point $(-a, 0)$, $\tan \theta = \pm \sqrt{\frac{3}{2}}$.

The values of $\tan \theta$ being all real and unequal, all of these points are double-points. The curve is shown in the figure.

REMARK. In the preceding theory of singular points it is assumed that the expression (2), § 97, can be developed in powers of ρ. If the function F is such that this development is impossible for certain values of x_0 and y_0, this impossibility may indicate a singular point at (x_0, y_0).

99. *Curve-tracing.* We have given rough figures of various curves in the preceding theory, and it is desirable that the student should know how to trace curves when their equations are given. The most elementary method is that of solving the equation for one co-ordinate, and then substituting various assumed values of the other co-ordinate in the solution, thus fixing various points of the curve. But unless the solution can be found by an equation of the first or second degree, this method will be tedious or impracticable. It may, however, commonly be simplified.

1. If the equation has no constant term, we may sometimes find the intersections of the curve with a number of lines through the origin. To do this we put

$$y = mx$$

in the equation, and then solve for x. The resulting values of x as a function of m are the abscissas of the points in which the curve cuts the line

$$y - mx = 0.$$

Then, by putting

$$m = \pm 1, \pm 2, \text{ etc.}; \quad m = \pm \tfrac{1}{2}, \pm \tfrac{1}{3}, \text{ etc.,}$$

we find as many points of intersection as we please.

To make this method practicable, the equations which we have to solve should not be of a degree higher than the second.

If the curve has a double-point, it may be convenient to take this point as the origin.

2. If the equation is symmetrical in x and y or x and $-y$, the curve will be symmetrical with respect to one of the lines $x - y = 0$ and $x + y = 0$.

The equation may then be simplified by referring it to new axes making an angle of 45° with the original ones.

The equations for transforming to such axes are
$$x = (x' + y') \sin 45°;$$
$$y = (x' - y') \sin 45°.$$

Application to the Folium of Descartes. If, in the equation of this curve,
$$x^3 + y^3 = 3axy,$$
we put $y = mx$, we shall find
$$x = \frac{3am}{1 + m^3}; \quad y = \frac{3am^2}{1 + m^3}.$$

We also find, from the equation of the curve and the preceding expressions for x and y in terms of m,
$$\frac{dy}{dx} = \frac{x^2 - ay}{ax - y^2} = \frac{2m - m^4}{1 - 2m^3}.$$

Then, for

$m = 1$, $x = \frac{3}{2}a$; $y = \frac{3}{2}a$; $\frac{dy}{dx} = -1$.

$m = 2$, $x = \frac{2}{3}a$; $y = \frac{4}{3}a$; $\frac{dy}{dx} = \frac{4}{5}$.

$m = \frac{3}{2}$, $x = \frac{36}{35}a$; $y = \frac{54}{35}a$; $\frac{dy}{dx} = \frac{33}{92}$.

$m = -2$, $x = \frac{6}{7}a$; $y = -\frac{12}{7}a$; $\frac{dy}{dx} = -\frac{20}{17}$.

etc. etc. etc. etc.

Thus we have, not only the points of the curve, but the tangents of the angle of direction of the curve at each point, which will assist us in tracing it.

CHAPTER XIV.

THEORY OF ENVELOPES.

100. The equation of a curve generally contains one or more constants, sometimes called *parameters*. For example, the equation of a circle,

$$(x - a)^2 + (y - b)^2 = r^2,$$

contains three parameters, a, b and r.

As another example, we know that the equation of a straight line contains two independent parameters.

Conceive now that the equation of any line, straight or curve, (which we shall call "the line" simply,) to be written in the implicit form

$$\phi(x, y, \alpha) = 0, \qquad (1)$$

α being a parameter. By assigning to α the several values α, α', α'', etc., we shall have an equal number of lines whose equations will be

$$\phi(x, y, \alpha) = 0; \quad \phi(x, y, \alpha') = 0; \quad \phi(x, y, \alpha'') = 0; \text{ etc.}$$

The collection of lines that can thus be formed by assigning all values to a parameter is called a **family of lines**.

Any two lines of the family, e.g., those which have α and α' as parameters, will in general have one or more points of intersection, determined by solving the corresponding equations for x and y. The co-ordinates, x and y, of the point of intersection will then come out as functions of α and α'.

Suppose the two parameters to approach infinitesimally near each other. The point of intersection will then approach a certain limit, which we investigate as follows:

Let us put
$$\alpha' = \alpha + \Delta\alpha.$$
The equations of the lines will then be
$$\phi(x, y, \alpha) = 0 \quad \text{and} \quad \phi(x, y, \alpha + \Delta\alpha) = 0.$$
If we develop the left-hand member of the second equation in powers of $\Delta\alpha$ by Taylor's theorem, it will become
$$\phi(x, y, \alpha) + \frac{d\phi}{d\alpha}\Delta\alpha + \frac{d^2\phi}{d\alpha^2}\frac{\Delta\alpha^2}{1\cdot 2} + \text{etc.} = 0.$$
Subtracting the first equation, dividing the remainder by $\Delta\alpha$, and passing to the limit, we find
$$\frac{d\phi(x, y, \alpha)}{d\alpha} = 0.$$
Hence the limit toward which the point (x, y) of intersection of two lines of a family approaches as the difference of the parameters becomes infinitesimal is found by determining x and y from the equations
$$\phi(x, y, \alpha) = 0 \quad \text{and} \quad \frac{d\phi(x, y, \alpha)}{d\alpha} = 0. \qquad (2)$$
The values of x and y thus determined will, in general, be functions of α; that is, we shall have
$$x = f_1(\alpha); \quad y = f_2(\alpha); \qquad (3)$$
which will give the values of the co-ordinates x and y of the limiting point of intersection for each value of α.

Now, suppose α to vary. Then x and y in (3) will also vary, and will determine a curve as the locus of x and y.

Such a curve is called the **envelope** of the family of lines, $\phi(x, y, \alpha) = 0$.

In (3) the equations of the curve are in the form of (2), § 76, α being the auxiliary variable. By eliminating α either from (2) or (3), we have an equation between x and y which will be the equation of the curve in the usual form.

101. Theorem. *The envelope and all the lines of the family which generate it are tangent to each other.*

Geometrically the truth of this will be seen by drawing a series of lines varying their position according to any continuous law, as in the first example of the following section. Taking three consecutive lines and numbering them (1), (2) and (3), it will be seen that as (1) and (3) approach (2) their points of intersection with (2) approach infinitely near each other. Since these infinitely near points of intersection also belong to the envelope, the line (2) passes through two infinitely near points of the envelope and is therefore a tangent to the envelope.

Analytic Proof. The equation of the envelope is found by eliminating α from the equations (2), and we may conceive this elimination to be effected by finding the value of α from the second of these equations (2), and substituting it in the first equation. That is, the equation

$$\phi(x, y, \alpha) = 0 \qquad (4)$$

represents any line of the original family when we regard α as a constant; and it represents the envelope when we regard α as a function of x and y, satisfying the equation

$$\frac{d\phi(x, y, \alpha)}{d\alpha} = 0. \qquad (5)$$

Let the value of α derived from this last equation be

$$\alpha = F(x, y). \qquad (6)$$

Now, to find the slope of the tangent to the original line of the family at the point (x, y), we differentiate (4), regarding α as a constant. Thus we have

$$\frac{d\phi}{dx} + \frac{d\phi}{dy}\frac{dy}{dx} = 0 \quad \text{or} \quad \frac{dy}{dx} = -\frac{D_x\phi}{D_y\phi}. \qquad (7)$$

If the original line is a straight one, this equation will give its slope.

To find the slope of the tangent to the envelope at the same point, we differentiate this same equation, regarding α as having the value (6). Thus we have

$$\frac{d\phi}{dx} + \frac{d\phi}{dy}\frac{dy}{dx} + \frac{d\phi}{d\alpha}\left(\frac{d\alpha}{dx} + \frac{d\alpha}{dy}\frac{dy}{dx}\right) = 0. \qquad (8)$$

But, because $\dfrac{d\phi}{d\alpha} = 0$, this equation will also give the value (7) for the slope; whence the curves have the same tangent at the point (x, y), and so are tangent to each other at this point.

102. We shall now illustrate this theory by some examples.

1. *To find the envelope of a straight line which moves so that the area of the triangle which it forms with the axes of coordinates is a constant.*

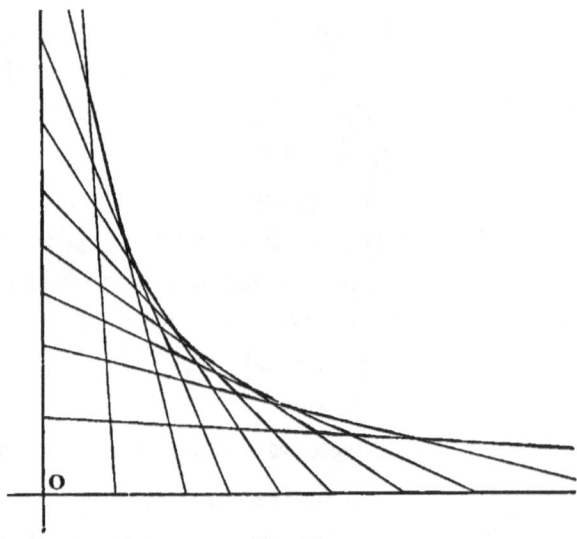

Fig. 39.

Since the area of the triangle is half the product of the intercepts of the axes cut off by the line, this product is also constant.

Calling a and b the intercepts, the equation of the line may be written in the form

$$\phi(x, y, a) = \frac{x}{a} + \frac{y}{b} - 1 = 0. \qquad (1)$$

THEORY OF ENVELOPES.

Here we have two varying parameters, a and b, while, to have an envelope, the change of the parameters must depend on a single varying quantity. But the condition that the product of the intercepts shall be constant enables us to eliminate one of the parameters, say b. We have, by this condition,

$$b = \frac{c}{a}, \qquad (2)$$

whence $\qquad \dfrac{db}{da} = -\dfrac{c}{a^2}.$

Now differentiating the equation (1) with respect to a, regarding b as a function of a, we have

$$\frac{d\phi}{da} = -\frac{x}{a^2} - \frac{y}{b^2}\frac{db}{da} = \frac{cy - b^2 x}{a^2 b^2} = \frac{y}{c} - \frac{x}{a^2} = 0. \qquad (3)$$

We have now to eliminate a from the equations (1) and (3), using (2) to eliminate b from (1). The easiest way to effect this elimination is as follows:

From (3) we have

$$a^2 y = cx; \quad a = \sqrt{\frac{cx}{y}}. \qquad (4)$$

Multiplying (1) by a, and substituting for b its value from (2), we have

$$x + \frac{a^2 y}{c} = a.$$

Substituting from (4), this equation becomes

$$2x = a = \sqrt{\frac{cx}{y}},$$

and thus the equation of the envelope becomes

$$xy = \tfrac{1}{4}c,$$

which is that of an hyperbola referred to its asymptotes.

This result coincides with one already found in Analytic Geometry, that tangents to an hyperbola cut off from the asymptotes intercepts whose product is a constant.

2. To find the envelope of the line for which the sum of the intercepts cut off from the co-ordinate axes is a constant.

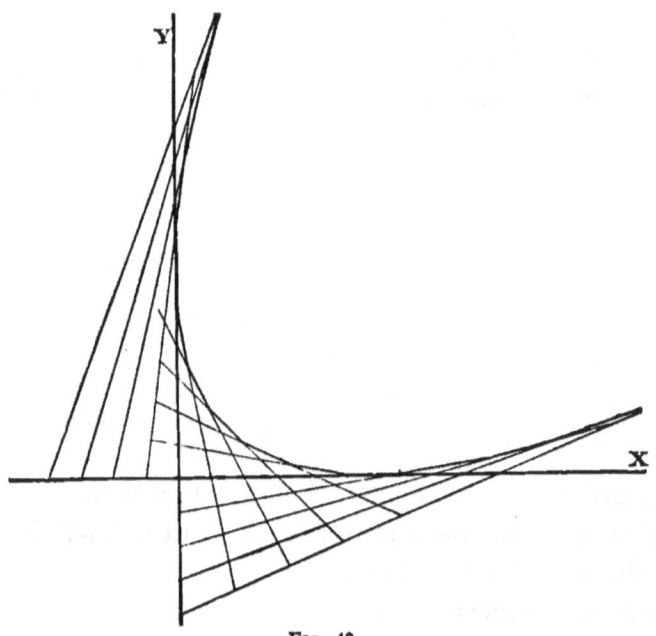

Fig. 40.

Let c be the constant sum of the intercepts. Then, if a be the one intercept, the other will be $c - a$. Thus the equation of the line is

$$\frac{x}{a} + \frac{y}{c-a} = 1,$$

in which a is the varying parameter.

Clearing of fractions, we may write the equation

$$\phi(x, y, a) = cx + a(y - x - c) + a^2 = 0,$$

whence
$$\frac{d\phi}{da} = y - x - c + 2a = 0.$$

From the last equation we have

$$a = \tfrac{1}{2}(x - y + c);$$

this value of a being substituted in the other gives

$$cx - \tfrac{1}{4}(x - y + c)^2 = 0,$$

or
$$(x - y)^2 - 2c(x + y) + c^2 = 0.$$

This equation, being of the second degree in the co-ordinates, is a *conic section*.

The terms of the second degree forming a perfect square, it is a *parabola*.

The equation of the axis of the parabola is

$$x - y = 0.$$

To find the two points in which the parabola cuts the axis of X we put $y = 0$, and find the corresponding values of x. The resulting equation is

$$x^2 - 2cx + c^2 = 0.$$

This is an equation with two equal roots, $x = c$, showing that the parabola touches the axis of X at the point $(c, 0)$. It is shown in the same way that the axis of Y is tangent to the parabola.

It may also be shown that the directrix and axis of the parabola each pass through the origin, and that the parameter is $\sqrt{2}c$.

3. If the difference of the intercepts cut off by a line from the axes is constant, it may be shown by a similar process that the envelope is still a parabola. This is left as an exercise for the student, who should be able to demonstrate the following results:

(α) When the sum of the intercepts is a positive constant, the parabola is in the first quadrant; when a negative constant, the parabola is in the third quadrant.

(β) When the difference, $a - b$, of the intercepts is a positive constant, the parabola is in the fourth quadrant; when a negative constant, in the second.

(γ) The co-ordinate axes touch the parabola at the ends of the parameter.

In each case the parabola touches each co-ordinate axis at a point determined by the value of the corresponding intercept when the other intercept vanishes, and each directrix intersects the origin at an angle of 45° with the axis.

4. Next take the case in which the sum of one intercept and a certain fraction or multiple of the other is a constant.

Let m be the fraction or multiplier. We then have

$$b + ma = c = \text{a constant.}$$

The equation of the line then becomes

$$\frac{x}{a} + \frac{y}{c - ma} - 1 = 0.$$

Proceeding as before, we find the equation of the envelope to be

$$(mx - y)^2 - 2c(mx + y) + c^2 = 0,$$

which is still the equation of a parabola.

5. *To find the envelope of a line which cuts off intercepts subject to the condition*

$$\frac{m^2}{a^2} + \frac{n^2}{b^2} = 1, \qquad (a)$$

m and n being constants.

We may simplify the work by substituting for the varying intercepts a and b the single variable parameter α determined by either of the equations

$$\sin \alpha \equiv \frac{m}{a}; \quad \cos \alpha \equiv \frac{n}{b}.$$

The equation of the varying line will then become

$$\phi(x, y) = \frac{x}{m} \sin \alpha + \frac{y}{n} \cos \alpha = 1. \qquad (1)$$

By differentiating with respect to α, we have

$$\frac{d\phi}{d\alpha} = \frac{x}{m} \cos \alpha - \frac{y}{n} \sin \alpha = 0. \qquad (2)$$

We may now eliminate α by simply taking the sum of the squares of these equations, which gives

$$\frac{x^2}{m^2} + \frac{y^2}{n^2} = 1,$$

the equation of an ellipse whose semi-axes are m and n.

THEORY OF ENVELOPES.

6. *To find the envelope of a circle of constant radius whose centre moves on a fixed circle.*

For convenience let us take the centre of the fixed circle as the origin, and put:

$a, b, \equiv$ the co-ordinates of the centre of the moving circle;
$c \equiv$ its radius;
$d \equiv$ the radius of the fixed circle.

The equation of the moving circle now becomes

$$(x-a)^2 + (y-b)^2 - c^2 = 0. \qquad (1)$$

By differentiation with respect to a,

$$x - a + (y - b)\frac{db}{da} = 0.$$

The condition that (a, b) lies on the fixed circle gives

$$a^2 + b^2 = d^2, \qquad (2)$$

whence $\qquad \dfrac{db}{da} = -\dfrac{a}{b}.$

Then, by substituting this value,

$$ay - bx = 0. \qquad (3)$$

We have now to eliminate a and b from (1), (2) and (3). Firstly, from (1) and (2), we find

$$x^2 + y^2 - 2ax - 2by = c^2 - d^2. \qquad (1')$$

From (2) and (3) we find the following expressions for a and b:

$$a = \frac{xd}{\sqrt{x^2 + y^2}}; \quad b = \frac{yd}{\sqrt{x^2 + y^2}}.$$

By substitution in (1'), and putting for brevity

$$r^2 \equiv x^2 + y^2,$$

we find $\qquad r^2 \pm 2rd + d^2 = c^2.$

Hence $\qquad r^2 = x^2 + y^2 = (c \pm d)^2,$

the equations of two circles around the origin as a centre, with radii $c + d$ and $c - d$.

7. Find the envelope of a family of ellipses referred to their centre and axes, the product of whose semi-axes is equal to a certain constant, c^2.

Ans. The equilateral hyperbola $xy = \tfrac{1}{2}c^2$.

8. *To find the envelope of a family of straight lines, such that the product of their distances from two fixed points is a constant.*

Let $(a, 0)$ and $(-a, 0)$ be taken as the two fixed points, and let c^2 be the constant. Also, let

$$x \cos \alpha + y \sin \alpha - p = 0 \qquad (1)$$

be the equation of any one of the lines in the normal form, p and α being the varying parameters.

The distances of the line from the points $(a, 0)$ and $(-a, 0)$ are respectively

$$-p + a \cos \alpha \quad \text{and} \quad -p - a \cos \alpha.$$

Hence we have the condition

$$p^2 - a^2 \cos^2 \alpha = c^2. \qquad (2)$$

Differentiating (1), regarding p as a function of α, we have

$$-x \sin \alpha + y \cos \alpha - \frac{dp}{d\alpha} = 0.$$

From (2) we obtain

$$\frac{dp}{d\alpha} = -\frac{a^2 \sin \alpha \cos \alpha}{p}.$$

We thus have the three equations

$$x \cos \alpha + y \sin \alpha = p, \qquad (a)$$

$$x \sin \alpha - y \cos \alpha = \frac{a^2 \sin \alpha \cos \alpha}{p}, \qquad (b)$$

$$p^2 = c^2 + a^2 \cos^2 \alpha$$
$$= c^2 + a^2 - a^2 \sin^2 \alpha, \qquad (c)$$

from which to eliminate p and α.

THEORY OF ENVELOPES. 179

To effect the elimination of α and p we find the values of x and y from (a) and (b) by taking

(a) $\times \cos \alpha + $ (b) $\times \sin \alpha$ and (a) $\times \sin \alpha - $ (b) $\times \cos \alpha$.

We thus find, by the aid of (c),

$$px = p^2 \cos \alpha + a^2 \sin^2 \alpha \cos \alpha;$$

$$x = (c^2 + a^2)\frac{\cos \alpha}{p};$$

$$y = c^2 \frac{\sin \alpha}{p}.$$

Hence
$$\frac{x}{c^2 + a^2} = \frac{\cos \alpha}{p};$$

$$\frac{y}{c^2} = \frac{\sin \alpha}{p}.$$

If we multiply the first of these equations by x and the second by y and add, then we have

$$\frac{x^2}{c^2 + a^2} + \frac{y^2}{c^2} = \frac{x \cos \alpha + y \sin \alpha}{p} = 1.$$

Hence the equation of the envelope is

$$\frac{x^2}{c^2 + a^2} + \frac{y^2}{c^2} = 1.$$

This represents an ellipse whose foci are the two fixed points.

This interpretation, however, presupposes that the product c^2 of the distances of the line from the two points is positive; that is, that the points are on the same side of the enveloping line. If the product is negative, the equation of the envelope will be

$$\frac{x^2}{a^2 - c^2} - \frac{y^2}{c^2} = 1,$$

which is the equation of an hyperbola.

These results give the theorem of Analytic Geometry that the product of the distances of a tangent from the foci of a conic is constant.

CHAPTER XV.

OF CURVATURE; EVOLUTES AND INVOLUTES.

103. *Position; Direction; Curvature.* The *position* of any point P on a curve is fixed by the values of the co-ordinates, x and y, of P. This is shown in Analytic Geometry.

If we have given, not only x and y, but the value of $\dfrac{dy}{dx}$ for the point P, then such value of the derivative indicates the *direction* of the curve at the point P, this direction being the same as that of the tangent at P.

The curve may also have a greater or less degree of *curvature* at P. The curvature is indicated by a change in the direction of the tangent, that is, in the value of $\dfrac{dy}{dx}$, when we pass to an adjacent point P'. But such change in the value of $\dfrac{dy}{dx}$ when we vary x is expressed by the value of the second derivative $\dfrac{d^2y}{dx^2}$. If this quantity is positive, the angle which the tangent makes with the axis of X is increasing with x at the point P, and the curve, viewed from below, is convex.

If $\dfrac{d^2y}{dx^2}$ is negative, the tangent is diminishing, and the curve, seen from below, is concave.

To sum up: If we take a value of the abscissa x, then the corresponding value of

y gives the *position* of a point P of the curve;

$\dfrac{dy}{dx}$ gives the *direction* of the curve at P;

$\dfrac{d^2y}{dx^2}$ depends upon the *curvature* of the curve at P.

CURVATURE; EVOLUTES AND INVOLUTES. 181

104. Contacts of Different Orders. Let two different curves be given by their respective equations:

$$y = f(x) \quad \text{and} \quad y = \phi(x).$$

If for a certain value of x, which value call x_0, the two values of y are equal, the two curves have the corresponding point in common; that is, they meet at the point (x_0, y).

If the values of $\dfrac{dy}{dx}$ are also equal at this point, it shows that the curves have the same direction at the point of meeting. They are then said to touch each other.

If the values of $\dfrac{d^2y}{dx^2}$ are also equal at this point, the two curves have also the same curvature at this point.

To show the result of these several equalities, let us give the abscissa x_0 (which we still take the same for both curves), an increment h, and develop the two values of y in powers of h by Taylor's theorem. To distinguish the values of y, $\dfrac{dy}{dx}$, etc., which belong to the two curves, we assign to one the suffix 0, and to the other the suffix 1. Then, for the one curve,

$$y = y_0 + \left(\frac{dy}{dx}\right)_0 h + \left(\frac{d^2y}{dx^2}\right)_0 \frac{h^2}{1 \cdot 2} + \cdots + \left(\frac{d^ny}{dx^n}\right)_0 \frac{h^n}{n!} + \text{etc.},$$

and, for the other,

$$y' = y_1 + \left(\frac{dy}{dx}\right)_1 h + \left(\frac{d^2y}{dx^2}\right)_1 \frac{h^2}{1 \cdot 2} + \cdots + \left(\frac{d^ny}{dx^n}\right)_1 \frac{h^n}{n!} + \text{etc.}$$

The difference between the values of y' and y is the intercept, between the two curves, of the ordinate at the point whose abscissa is $x_0 + h$. Its expression is

$$y_1 - y_0 + \left[\left(\frac{dy}{dx}\right)_1 - \left(\frac{dy}{dx}\right)_0\right]h + \left[\left(\frac{d^2y}{dx^2}\right)_1 - \left(\frac{d^2y}{dx^2}\right)_0\right]\frac{h^2}{1 \cdot 2} + \text{etc.}$$

Now, consider the case in which the curves meet at the point P, whose abscissa is x_0. Then

$$y_1 - y_0 = 0,$$

and the intercept of the ordinate will be

$$\left[\left(\frac{dy}{dx}\right)_1 - \left(\frac{dy}{dx}\right)_0\right]h + \text{terms in } h^2, \text{ etc.},$$

which, when h becomes infinitesimal, is an infinitesimal of the first order.

If we also have

$$\left(\frac{dy}{dx}\right)_1 = \left(\frac{dy}{dx}\right)_0,$$

the ordinates will differ only by a quantity containing h^2 as a factor, and so of the second order. Hence:

When two curves are tangent to each other, they are separated only by quantities of at least the second order at an infinitesimal distance from the point of tangency.

In the same way it is shown that if the second differential coefficient also vanishes, the separation will be of the third order, and so on.

Def. When two curves are tangent to each other, if the first n differential coefficients for the two curves are equal at the point of tangency, the curves are said to have **contact of the nth order.**

Hence a case of simple tangency is a contact of the first order. If the second derivatives are also equal, the contact is of the second order, and so forth.

105. THEOREM. *In contacts of an even order the two curves intersect at the point of contact; in those of an odd order they do not.*

For, in contact of the nth order, the first term of $y' - y$ (§ 104) which does not vanish contains h^{n+1} as a factor.

If n is odd, $n + 1$ is even, and $y' - y$ has the same algebraic sign whether we take h positively or negatively. Hence the curves do not intersect.

If n is even, $n + 1$ is odd, and the values of $y' - y$ have opposite signs on the two sides of the point of contact, thus showing that the curves intersect.

CURVATURE; EVOLUTES AND INVOLUTES. 183

106. *Radius of Curvature.* The curvature at any point is measured thus: We pass from the point P to a point P' infinitesimally near it. The curvature is then measured by the ratio of the change in the direction of the tangent (or normal) to the distance PP'. Let us put

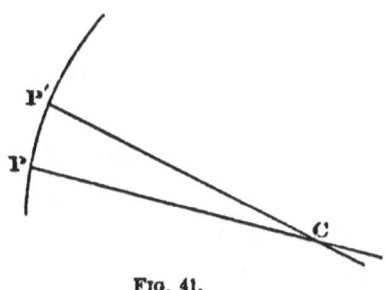

Fig. 41.

$\alpha \equiv$ the angle which the tangent at P makes with the axis of X.

$\alpha + d\alpha \equiv$ the same angle for the tangent at P'.

$ds \equiv$ the infinitesimal distance PP'.

Then, by definition,

$$\text{Curvature} = \frac{d\alpha}{ds}.$$

Now, because $\tan \alpha = \dfrac{dy}{dx}$,

we have, by differentiation,

$$\sec^2 \alpha \, d\alpha = \frac{d^2y}{dx^2} dx.$$

Also, $\sec^2 \alpha = 1 + \tan^2 \alpha = 1 + \dfrac{dy^2}{dx^2}$

and $ds = \sqrt{\left(1 + \dfrac{dy^2}{dx^2}\right)} dx.$

From these equations we readily derive

$$\text{Curvature} = \frac{d\alpha}{ds} = \frac{\dfrac{d^2y}{dx^2}}{\left(1 + \dfrac{dy^2}{dx^2}\right)^{\frac{3}{2}}}.$$

Now, draw normals to the curve at the points P and P', and let C be their point of intersection. Because they are perpendicular to the tangents, the angle PCP' between them will be $d\alpha$, and if we put

$$\rho = PC,$$

we shall have
$$PP' = ds = \rho d\alpha.$$

Hence $\quad \rho = \dfrac{ds}{d\alpha} = \dfrac{1}{\text{curvature}} = \dfrac{\left(1 + \dfrac{dy^2}{dx^2}\right)^{\frac{3}{2}}}{\dfrac{d^2y}{dx^2}}.$

The length ρ is called the **radius of curvature** at the point P, and C is called the **centre of curvature**.

COROLLARY. *The centre of curvature for any point of a curve is the intersection of consecutive normals cutting the curve infinitely near that point.*

107. *The Osculating Circle.* If, on the normal PC to any curve at the point P, we take any point as the centre of a circle through P, that circle will be tangent to the curve at P; that is, it will, in general, have contact of the first order at P. But there is one such circle which has contact of a higher order, namely, that whose centre is at the centre of curvature. Since this circle will have the same curvature at P as the curve itself has, it will have contact of at least the second order at P.

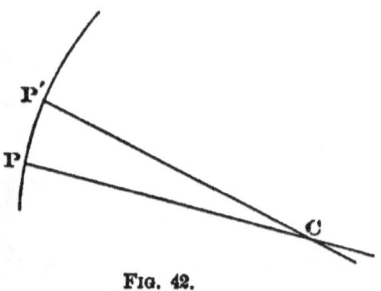

Fig. 42.

This proposition is rigorously demonstrated by finding that circle which shall have contact of the second order with the curve at the point P.

Let us put

x, y, the co-ordinates of P;

$p \equiv \dfrac{dy}{dx}$ for the curve at the point P;

$q \equiv \dfrac{d^2y}{dx^2}$ for the curve at the point P.

CURVATURE; EVOLUTES AND INVOLUTES. 185

These last two quantities are found by differentiating the equation of the curve.

Now, $\dfrac{dy}{dx}$ and $\dfrac{d^2y}{dx^2}$ must have these same values at the point (x, y) in the case of the circle having contact of the second order (§ 104).

Let the equation of this circle be

$$(x - a)^2 + (y - b)^2 = r^2. \qquad (a)$$

By differentiation, we have

$$(x - a)dx + (y - b)dy = 0,$$

whence
$$\frac{dy}{dx} = \frac{x - a}{b - y} = p. \qquad (b)$$

Differentiating again,

$$\frac{d^2y}{dx^2} = \frac{1}{b - y} + \frac{(x - a)}{(b - y)^2}\frac{dy}{dx} = \frac{(y - b)^2 + (x - a)^2}{(y - b)^3}$$

$$= \frac{r^2}{(y - b)^3} = q. \qquad (c)$$

From (b) combined with (a) we find

$$1 + p^2 = 1 + \frac{(x - a)^2}{(y - b)^2} = \frac{r^2}{(y - b)^2};$$

$$\therefore (1 + p^2)^{\frac{3}{2}} = \frac{r^3}{(y - b)^3}.$$

Dividing this by (c) gives

$$r = \frac{(1 + p^2)^{\frac{3}{2}}}{q},$$

the equivalent of the expression already found for the radius of curvature.

Hence if we determine a circle by the condition that it shall have contact of the second order with the curve at the point P, its radius will be equal to the radius of curvature. This circle is called the **osculating circle** for the point P. Each point of a curve has its osculating circle, determined by the position, direction and curvature at that point.

Cor. The osculating circle will, in general, intersect the curve at the point of contact, for it has contact of the second order.

This may also be seen by reflecting that the curvature of a curve is, in general, a continuously varying quantity as we pass along the curve, and that, at the point of contact, it is equal to the curvature of the circle. Hence, on one side of the point of contact, the curvature of the curve is less than that of the circle, and so the curve passes without the circle; and on the other side the curvature of the curve is greater, and thus the curve passes within the circle.

If, however, the curvature should be a maximum or a minimum at the point of contact, it will either increase on both sides of this point or diminish on both sides, whence the circle will not intersect the curve.

✗108. *Radius of Curvature when the Abscissa is not taken as the Independent Variable.* Suppose that, instead of x, some other variable, u, is regarded as the independent variable. We then have

$$y = f_1(u); \quad x = f_2(u).$$

Now, it has been shown that, in this case, we have (§ 56)

$$\frac{d^2y}{dx^2} = \frac{\dfrac{d^2y}{du^2}\dfrac{dx}{du} - \dfrac{d^2x}{du^2}\dfrac{dy}{du}}{\left(\dfrac{dx}{du}\right)^3}. \quad (2)$$

Also, we have

$$1 + \left(\frac{dy}{dx}\right)^2 = 1 + \frac{\left(\dfrac{dy}{du}\right)^2}{\left(\dfrac{dx}{du}\right)^2} = \frac{\left(\dfrac{dx}{du}\right)^2 + \left(\dfrac{dy}{du}\right)^2}{\left(\dfrac{dx}{du}\right)^2}. \quad (3)$$

These expressions being substituted in the expression for the radius of curvature, it becomes

$$\rho = \frac{\left\{\left(\dfrac{dx}{du}\right)^2 + \left(\dfrac{dy}{du}\right)^2\right\}^{\frac{3}{2}}}{\dfrac{d^2y}{du^2}\dfrac{dx}{du} - \dfrac{d^2x}{du^2}\dfrac{dy}{du}}. \qquad (4)$$

109. *Radius of Curvature of a Curve referred to Polar Co-ordinates.* Let the equation of the curve be given in the form

$$r = \phi(\theta).$$

The preceding expression (4) may be employed in this case by taking the angle θ as the independent variable. By differentiating the expressions

$$x = r \cos \theta,$$
$$y = r \sin \theta,$$

regarding r as a function of θ, we find, when we put, for brevity,

$$r' \equiv \frac{dr}{d\theta}, \quad r'' = \frac{d^2r}{d\theta^2},$$

$$\frac{dx}{d\theta} = -r \sin \theta + r' \cos \theta;$$

$$\frac{d^2x}{d\theta^2} = (r'' - r)\cos \theta - 2r' \sin \theta;$$

$$\frac{dy}{d\theta} = r \cos \theta + r' \sin \theta;$$

$$\frac{d^2y}{d\theta^2} = (r'' - r)\sin \theta + 2r' \cos \theta.$$

By substituting these derivatives with respect to θ for those with respect to u in (4) and performing easy reductions, we find

$$\rho = \frac{(r^2 + r'^2)^{\frac{3}{2}}}{r^2 - rr'' + 2r'^2} = \frac{\left\{r^2 + \left(\dfrac{dr}{d\theta}\right)^2\right\}^{\frac{3}{2}}}{r^2 - r\dfrac{d^2r}{d\theta^2} + 2\left(\dfrac{dr}{d\theta}\right)^2}. \qquad (5)$$

which is the required expression for the radius of curvature.

EXAMPLES AND EXERCISES.

1. *The Parabola.* To find the radius of curvature of a curve at any point, we have to form the value of ρ from the equation of the curve. The equation of the parabola is

$$y^2 = 2px,$$

whence we find

$$\frac{dy}{dx} = \frac{p}{y};$$

$$\frac{d^2y}{dx^2} = -\frac{p^2}{y^3}.$$

Then, by substituting in the expression for ρ, we find

$$\rho = \frac{(y^2 + p^2)^{\frac{3}{2}}}{p^2},$$

the negative sign being omitted, because we have no occasion to apply any sign to ρ.

At the vertex $y = 0$, whence

$$\rho = p.$$

Hence, at the vertex, the radius of curvature is equal to the semi-parameter, and the centre of curvature is therefore twice as far from the vertex as the focus is.

2. Show that the radius of curvature at any point (x, y) of an ellipse is

$$\rho = \frac{(a^4y^2 + b^4x^2)^{\frac{3}{2}}}{a^4b^4},$$

and show that at the extremities of the axes it is a third proportional to the semi-axes.

3. Show that the algebraic expression for ρ is the same in the case of the hyperbola as in that of the ellipse.

4. What must be the eccentricity of an ellipse that the centre of curvature for a point at one end of the minor axis may lie on the other end of that axis? *Ans.* $e = \sqrt{\frac{1}{2}}$.

5. Show that in the case supposed in the last problem the radius of curvature at an end of the major axis will be one fourth that axis.

6. *The Cycloid.* By differentiating the equations (1), § 80, of the cycloid, we find

$$\left.\begin{array}{r}\dfrac{dx}{du} = a - a\cos u = y; \\[4pt] \dfrac{d^2x}{du^2} = \dfrac{dy}{du} = a\sin u; \\[4pt] \dfrac{d^2y}{du^2} = a\cos u. \end{array}\right\} \quad (2)$$

Then, by substituting in (4) and reducing, we find, for the radius of curvature,

$$\rho = 2^{\frac{3}{2}}a\sqrt{1 - \cos u} = 4a\sin \tfrac{1}{2}u.$$

We see that at the cusp, O, of the cycloid, where $u = 0$, the radius of curvature also becomes zero.

7. *The Archimedean Spiral.* Show from (5) that the radius of curvature of this spiral ($r = a\theta$) is

$$\rho = \frac{a(1 + \theta^2)^{\frac{3}{2}}}{2 + \theta^2}.$$

8. *The Logarithmic Spiral.* The equation of the logarithmic spiral being

$$r = ae^{l\theta},$$

show that the radius of curvature is

$$\rho = r\sqrt{1 + l^2}.$$

Hence show that the line drawn from the centre of curvature of any point P of the spiral to the pole is perpendicular to the radius vector of the point P.

9. Show that the radius of curvature of the lemniscate in terms of polar co-ordinates is

$$\rho = \frac{a}{3\sqrt{\cos 2\theta}} = \frac{a^2}{3r}.$$

110. *Evolutes and Involutes.* For every point of a curve there is a centre of curvature, found by the preceding formulæ. The locus of all such centres is called the **evolute** of the curve.

To find the evolute of a curve, let $(x_{\prime}, y_{\prime})$ be the co-ordinates of any point P of the curve; PC, the radius of curvature for this point; and α, the angle which the tangent at P makes with the axis of X. Then, for the co-ordinates of C, we have

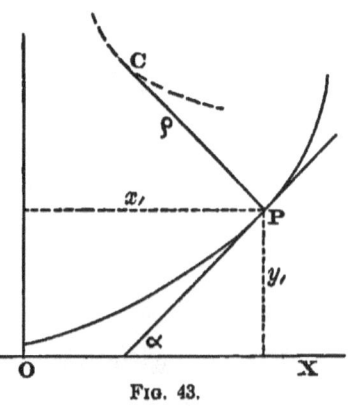

Fig. 43.

$$x = x_{\prime} - \rho \sin \alpha;$$
$$y = y_{\prime} + \rho \cos \alpha.$$

Substituting for ρ its value (§ 106), and for $\sin \alpha$ and $\cos \alpha$ their values from the equation

$$\tan \alpha = \frac{dy_{\prime}}{dx_{\prime}},$$

we find

$$\left.\begin{array}{l} x = x_{\prime} - \dfrac{1 + \dfrac{dy_{\prime}^{2}}{dx_{\prime}^{2}}}{\dfrac{d^{2}y_{\prime}}{dx_{\prime}^{2}}} \dfrac{dy_{\prime}}{dx_{\prime}}; \\[2em] y = y_{\prime} + \dfrac{1 + \dfrac{dy_{\prime}^{2}}{dx_{\prime}^{2}}}{\dfrac{d^{2}y_{\prime}}{dx_{\prime}^{2}}}. \end{array}\right\} \quad (1)$$

If in the second members of these equations we substitute the values of the derivatives obtained from the equation of the curve, we shall have two equations between the four variables x, y, $x_{\prime}$ and $y_{\prime}$. By eliminating $x_{\prime}$ and $y_{\prime}$ from these equations and that of the given curve, we shall have a single equation between x and y, which will be that of the evolute.

111. *Case of an Auxiliary Variable.* If the equation of the curve is expressed by an auxiliary variable, we have to make in (1) the same substitution of the values of $\dfrac{dy_{\prime}}{du}$, $\dfrac{d^2 y_{\prime}}{du^2}$, etc., as in § 108. Thus we find, instead of (1),

$$\left. \begin{array}{l} x = x_{\prime} - \dfrac{dy_{\prime}}{du} \dfrac{\left(\dfrac{dx_{\prime}}{du}\right)^2 + \left(\dfrac{dy_{\prime}}{du}\right)^2}{\dfrac{d^2 y_{\prime}}{du^2}\dfrac{dx_{\prime}}{du} - \dfrac{d^2 x_{\prime}}{du^2}\dfrac{dy_{\prime}}{du}}; \\[2em] y = y_{\prime} + \dfrac{dx_{\prime}}{du} \dfrac{\left(\dfrac{dx_{\prime}}{du}\right)^2 + \left(\dfrac{dy_{\prime}}{du}\right)^2}{\dfrac{d^2 y_{\prime}}{du^2}\dfrac{dx_{\prime}}{du} - \dfrac{d^2 x_{\prime}}{du^2}\dfrac{dy_{\prime}}{du}}; \end{array} \right\} \quad (2)$$

which are the equations of the evolute in the same form.

EXAMPLES OF EVOLUTES.

112. *The Evolute of the Parabola.* If we substitute in (1) for the derivatives of $y_{\prime}$ with respect to $x_{\prime}$ the values already found for the parabola, these equations (1) become

$$x = x_{\prime} + p + \frac{y_{\prime}^2}{p} = p + \frac{3}{2}\frac{y_{\prime}^2}{p};$$

$$y = -\frac{y_{\prime}^3}{p^2}.$$

We now have to eliminate $y_{\prime}$ from these two equations, $x_{\prime}$ having already been eliminated by the equation of the curve. They give

$$y_{\prime}^2 = \tfrac{2}{3}p(x - p); \quad y_{\prime}^3 = -p^2 y.$$

Equating the cube of the first equation to the square of the second, we find, for the equation of the evolute of the parabola,

$$y^2 = \frac{8}{27}\frac{(x-p)^3}{p}.$$

113. *Evolute of the Ellipse.* From the equation of the ellipse, we find

$$\frac{dy_{\prime}}{dx_{\prime}} = -\frac{b^2 x_{\prime}}{a^2 y_{\prime}}; \quad \frac{d^2 y_{\prime}}{dx_{\prime}^2} = -\frac{b^4}{a^2 y_{\prime}^3}.$$

By substituting in (1) and reducing, we find

$$x = x_{\prime} - \frac{x_{\prime} y_{\prime}^2}{b^2}\left(1 + \frac{b^4 x_{\prime}^2}{a^4 y_{\prime}^2}\right) = x_{\prime}\frac{a^4 b^2 - a^4 y_{\prime}^2 - b^4 x_{\prime}^2}{a^4 b^2}.$$

Remarking that the equation of the ellipse gives

$$a^4 b^2 - a^4 y_{\prime}^2 = a^2 b^2 x_{\prime}^2,$$

and putting $\quad e^2 \equiv a^2 - b^2,$

the preceding equation becomes

$$x = \frac{c^2 x_{\prime}^3}{a^4}. \tag{a}$$

In the same way we get

$$y = y_{\prime} - \frac{a^2 y_{\prime}^2}{b^4}\left(1 + \frac{b^4 x_{\prime}^2}{a^4 y_{\prime}^2}\right) = -\frac{c^2 y_{\prime}^3}{b^4}. \tag{b}$$

In this case the easiest way to effect the elimination of $x_{\prime}$ and $y_{\prime}$ is to obtain the values of these quantities from (a) and (b), and then substitute them in the equation of the ellipse. From (a) and (b), we find

$$x_{\prime} = \left(\frac{a^4 x}{c^2}\right)^{\frac{1}{3}}; \quad y_{\prime} = -\left(\frac{b^4 y}{c^2}\right)^{\frac{1}{3}},$$

which values are to be substituted in the equation

$$\frac{x_{\prime}^2}{a^2} + \frac{y_{\prime}^2}{b^2} = 1.$$

We thus find, for the equation of the evolute of the ellipse,

$$a^{\frac{2}{3}} x^{\frac{2}{3}} + b^{\frac{2}{3}} y^{\frac{2}{3}} = c^{\frac{2}{3}}.$$

The figure shows the form of the curve. The following properties should be deduced by the student.

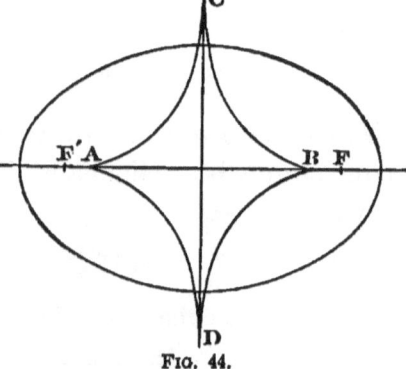

Fig. 44.

(a) The evolute lies wholly within the ellipse, or cuts it (as in the figure), according as $e^2 < \frac{1}{2}$ or $e^2 > \frac{1}{2}$.

(b) The ratio $CD : AB$ (which lines we may call axes of the evolute) is the inverse of the ratio of the corresponding axes of the ellipse.

114. *Evolute of the Cycloid.* Here we have to apply the formulæ (2) for the case of a separate independent variable. Substituting in (2) the values of the derivatives already given for the cycloid, we shall find

$$\left(\frac{dx}{du}\right)^2 + \left(\frac{dy}{du}\right)^2 = 2a^2(1 - \cos u);$$

$$\frac{d^2y}{du^2}\frac{dx}{du} - \frac{d^2x}{du^2}\frac{dy}{du} = -a^2(1 - \cos u);$$

$$x = x_1 + 2a \sin u = a(u + \sin u);$$
$$y = y_1 - 2a(1 - \cos u) = -a(1 - \cos u).$$

These last two equations are those of the evolute.

Let us investigate its form. For $u = 0$ we have $x = 0$ and $y = 0$, whence the origin is a point of the curve.

For $u = \pi$ we have

$x = a\pi;$
$y = -2a;$

giving a point C, below the middle of the base of the cycloid, at the dis-

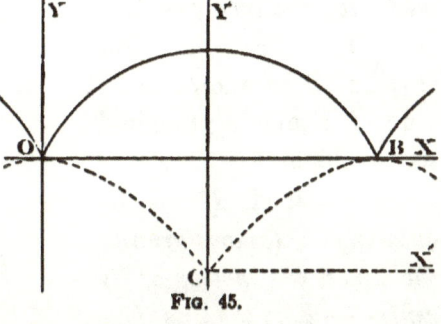

Fig. 45.

tance $2a$. Let us take this point as a new origin, and call the co-ordinates referred to it x' and y'. We then have

$$x' = x - a\pi = a(\theta - \pi + \sin \theta);$$
$$y' = y + 2a = a(1 + \cos \theta).$$

If we now put

$$\theta' \equiv \theta - \pi,$$

these equations become

$$x' = a(\theta' - \sin \theta');$$
$$y' = a(1 - \cos \theta');$$

which are the equations of another cycloid, equal to the original one, and similarly situated. The cycloid therefore posesses the remarkable property of being identical in form with its own evolute.

115. *Fundamental Properties of the Evolute.*

THEOREM I. *The involute of a curve is the envelope of its normals.*

As we move along a curve, the normal will be a straight line moving according to a certain law depending upon the form of the curve. This line will, in general, have an envelope, which envelope will be, by definition, the locus of the point of intersection of consecutive normals. But this point has been shown to be the centre of curvature, whose locus is, by definition, the evolute. Hence follows the theorem.

COROLLARY. *The normals to a curve are tangents to its evolute.* For this has been shown to be true of a moving line and its envelope.

THEOREM II. *If the osculating circle move around the curve, the motion of its centre is along the line joining that centre to the point of contact.*

This theorem will be made evident by a study of the figure. If the line $P_2 C_2$ be one of the normals from the point of contact P_2 to the centre, then, since

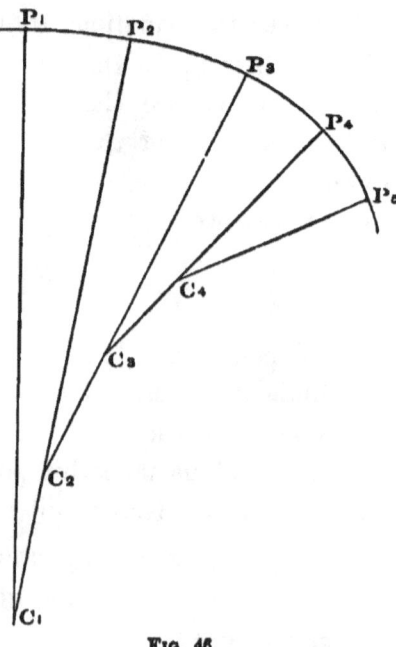

Fig. 46.

CURVATURE; EVOLUTES AND INVOLUTES. 195

this normal is tangent to the locus of the centre, it will be the line along which the centre is moving at the instant.

THEOREM III. *The arc of the evolute contained between any two points is equal to the difference of the radii of the osculating circles whose centres are at these points.*

For, if we suppose the points C_1, C_2, etc., to approach infinitesimally near each other, then, since the infinitesimal arcs C_1C_2, C_2C_3, etc., are coincident with those successive radii of the osculating circle which are normal to the curve, these radii are continually diminished by these same infinitesimal amounts.

The analytic proof of Theorems II. and III. is as follows: Let the equation of the osculating circle be

$$(x - a)^2 + (y - b)^2 = \rho^2,$$

where a and b are the co-ordinates of the centre of curvature, and therefore of a point of the evolute.

The complete differential of this equation gives

$$(x - a)(dx - da) + (y - b)(dy - db) = \rho d\rho. \quad (a)$$

If, in this equation, we suppose x and y to be the co-ordinates of the point of contact of the circle with the curve, then dx and dy will have the same value at this point whether we conceive them to belong to the circle, supposed for the moment to be fixed, or to the curve. But in the fixed circle we have

$$(x - a)dx + (y - b)dy = 0. \quad (b)$$

Subtracting this equation from (a) and dividing by ρ, we find

$$\frac{x - a}{\rho}da + \frac{y - b}{\rho}db = - d\rho, \quad (c)$$

which is a relation between the differential of the co-ordinates of the centre and the differential of the radius. Now, if we put β for the angle which the normal radius makes with the axis of X, we have

$$\frac{x - a}{\rho} = \cos \beta; \quad \frac{y - b}{\rho} = \sin \beta. \quad (d)$$

But this same normal radius is a tangent to the evolute. If we call σ the arc of the evolute, we find by a simple construction $da = \cos\beta d\sigma$; $db = \sin\beta d\sigma$.

Multiplying these equations by $\cos\beta$ and $\sin\beta$, respectively, and adding, we find

$$d\sigma = \cos\beta da + \sin\beta db.$$

Comparing (c) and (d), we find

$$d\sigma = -d\rho,$$
or
$$d(\sigma + \rho) = 0.$$

Now, a quantity whose differential is zero is a constant. Hence we always have

$$\sigma + \rho = \text{constant},$$
or
$$\sigma = \text{constant} - \rho.$$

If we represent by σ_1 and σ_2 the arcs from any arbitrary point of the involute to the two chosen points, and by ρ_1 and ρ_2 the values of ρ for these points, we have

$$\sigma_1 = \text{const.} - \rho_1;$$
$$\sigma_2 = \text{const.} - \rho_2.$$
$$\therefore \sigma_2 - \sigma_1 = \rho_1 - \rho_2,$$

or the intercepted arc equal to the difference of the radii, as was to be proved.

It must be remarked, however, that whenever we pass a cusp on the evolute, we must regard the arc as negative on one side and positive on the other. In the case of the ellipse, for example, those radii will be equal which terminate at equal distances on the two sides of any cusp, as A, B, C or D, and the intercepted arc must then be taken as zero.

116. *Involutes.* The **involute** of a curve C is that curve which has C as its evolute.

The fundamental property of the involute is this: The involute may be formed from the evolute by rolling a tangent line upon the latter. A point P on the rolling tangent will then describe the involute.

This will be seen by reference to Fig. 46. The rolling line, being tangent to the evolute, coincides with the radius P_1C_1, and as it rolls along the evolute into successive positions, P_2C_2, P_3C_3, etc., the motion of the point P is continually normal to its direction.

It will also be seen that the radius of curvature of the involute at each point is equal to the distance PC from P to the point of contact with the evolute.

The conception may be made clearer by conceiving the rolling line to be represented by a string which is wrapped around the evolute. The involute is then formed by the motion of a point on the string.

The general method of determining the involutes of given curves involves the integral calculus.

PART II.

THE INTEGRAL CALCULUS.

PART II.
THE INTEGRAL CALCULUS.

CHAPTER I.
THE ELEMENTARY FORMS OF INTEGRATION.

117. *Definition of Integration.* Whenever we have given function of a variable x, say

$$u = F(x),$$

e may, by differentiation, obtain another function of x,

$$\frac{du}{dx} = F'(x),$$

hich we call the *derived function*.

In the integral calculus we consider the reverse process. ʃe have given a derived function

$$F'(x),$$

ıd the problem is: *What function or functions, when differ- ıtiated, will have $F'(x)$ as their derivative?*

Every such function is called an **integral** of $F'(x)$.

The process of finding the integral is called **integration**.

The operation of integration is indicated by the sign $\int$, ılled "integral of," written before the product of the given ınction by the differential of the variable. Thus the ex- ression

$$\int F'(x)dx$$

ıeans: *that function whose differential with respect to x is $"(x)dx$.*

Calling u the required function, then if we have
$$\frac{du}{dx} = F'(x),$$
we must also have
$$u = \int F'(x)dx = \int \frac{du}{dx}dx.$$

As examples:

Because $\qquad d(x^2) = 2xdx,$

we have $\qquad \int 2xdx = x^2.$

Because $\quad d(ax^2 + bx + c) = (2ax + b)dx,$

we have $\qquad \int (2ax + b)dx = ax^2 + bx + c.$

And, in general, if, by differentiation, we have
$$dF(x) = F'(x)dx,$$
we shall have $\qquad \int F'(x)dx = F(x).$

118. *Arbitrary Constant of Integration.* The followin principle is a fundamental one of the integral calculus:

If $F(x)$ is the integral of any derived function of the va riable x, then every function of the form
$$F(x) + h,$$
h being any quantity whatever independent of x, will also l an integral.

This follows immediately from the fact that h will dis appear in differentiation, so that the two functions
$$F(x) \quad \text{and} \quad F(x) + h$$
have the same derivative (cf. § 24).

The same principle may be seen from another point c view: Since the problem of differentiation is to find a func tion which, being differentiated, will give a certain resull and since any quantity independent of the variable whic may be added to the original function will have disappeare by differentiation, it follows that we must, to have the mos general expression for the integral, add this possible but unknown quantity to the integral.

The quantity thus added is called an *arbitrary constant*. But it must be well understood that the word *constant* merely means *independent of the variable with reference to which the integration is performed*.

It follows from all this that the integral can never be completely found from the differential equation alone, but that some other datum is needed to determine the arbitrary constant and thus to complete the solution.

Such a datum is the value of the integral for some one value of the variable. Let $F(x) + h$ be the integral, and let it be given that

when $x = a$, then the integral $= K$.

We must have, by this datum,

$$F(a) + h = K,$$

which gives $\qquad h = K - F(a),$

and thus determines h.

REMARK. Any symbol may be taken to represent the arbitrary constant. The letters c and h are those most generally used. We may affix to it either the positive or the negative sign, and may represent it by any function of arbitrary but constant quantities which we find it convenient to introduce. It is often advantageous to write it as a quantity of the same kind as the variable which is integrated.

119. *Integration of Entire Functions.*

THEOREM I. *The integral of any power of a variable is the power higher by unity, divided by the increased exponent.*

In symbolic language, we have

$$\int x^n dx = \frac{x^{n+1}}{n+1} + h.$$

For, by differentiating the expression $\frac{x^{n+1}}{n+1} + h$, we have x^n.

THEOREM II. *Any constant factor of the given differential may be written before the sign of integration.*

In symbolic language,

$$\int aF'(x)dx = a\int F'(x)dx.$$

This is the converse of the Theorem of § 23. By that theorem we have

$$d(aF(x)) = adF(x),$$

from which the above converse theorem at once follows.

In the special case $a = -1$ we have

$$\int -F'(x)dx = \int F'(x)d(-x) = -\int F'(x)dx.$$

Hence the corollary: *If the integral is preceded by the negative sign we may place that sign before either the derived function or the differential.*

THEOREM III. *If the derived function is a sum of several terms, the integral is the sum of the separate integrals of the terms.*

In symbolic language,

$$\int (X + Y + Z + \ldots)dx = \int Xdx + \int Ydx + \int Zdx + \ldots$$

This, again, is the converse of Theorem II of § 22.

The foregoing theorems will enable us to find the integral of any entire function of a variable. To take the function in its most general form, let it be required to find the integral

$$u = \int (ax^m + bx^n + cx^p + \ldots)dx.$$

By Theorem III.,

$$u = \int ax^m dx + \int bx^n dx + \int cx^p dx + \ldots$$

By Theorem II.,
$$\int ax^m dx = a \int x^m dx;$$
etc.   etc.;

and by Theorem I.,
$$\int x^m dx = \frac{x^{m+1}}{m+1} + h_1;$$
etc.   etc.

By successive substitution we then have
$$u = \frac{ax^{m+1}}{m+1} + \frac{bx^{n+1}}{n+1} + \frac{cx^{p+1}}{p+1} + ah_1 + bh_2 + ch_3 + \ldots,$$
where h_1, h_2, h_3, etc., are the arbitrary constants added to the separate integrals.

Since the sum of the products of any number of constants by constant factors is itself a constant, we may represent the sum $ah_1 + bh_2 + ch_3$ by the single symbol h. Thus we have
$$\int (ax^m + bx^n + cx^p + \ldots) dx$$
$$= \frac{ax^{m+1}}{m+1} + \frac{bx^{n+1}}{n+1} + \frac{cx^{p+1}}{p+1} + \ldots + h.$$

EXERCISES.

Form the integrals of the following expressions, multiplied by dx:

1. x^2.
2. x^3.
3. x^{-2}.
4. x^{-3}.
5. ax^2.
6. bx^3.
7. ax^{-2}.
8. bx^{-3}.
9. $ax + b$.
10. $ax^2 - c$.
11. $ax^2 + cx$.
12. $ax^3 - cx^2$.
13. $x^{\frac{1}{2}}$.
14. $x^{\frac{3}{2}}$.
15. $x^{-\frac{1}{2}}$.
16. $ax^{-\frac{1}{2}}$.
17. $ax^{\frac{1}{2}} - bx^{-\frac{1}{2}}$.
18. $mx^{\frac{1}{2}} - \frac{n}{x^{\frac{1}{2}}}$.
19. $\frac{a}{x^2} - \frac{b}{x^3}$.
20. $a + \frac{1}{x^2}$.

120. *The Logarithmic Function.* An exceptional case of Theorem I. occurs when $n = -1$, because then $n + 1 = 0$, and the function becomes infinite in form. But since
$$d \cdot \log x = \frac{dx}{x} = x^{-1} dx,$$

it follows that we have for this special case

$$\int x^{-1}dx = \int \frac{dx}{x} = \log x + h. \qquad (a)$$

Let c be the number of which h is the logarithm. We then have
$$\log x + h = \log x + \log c = \log cx.$$

We may equally suppose
$$h = -\log c = \log \frac{1}{c}.$$

Then
$$\log x + h = \log \frac{x}{c}.$$

Hence we may write either

$$\left.\begin{array}{l} \int \dfrac{dx}{x} = \log cx, \\[1ex] \int \dfrac{dx}{x} = \log \dfrac{x}{c}; \end{array}\right\} \qquad (b)$$

or

c being an arbitrary constant.

We thus have the principle: *The arbitrary constant added to a logarithm may be introduced by multiplying or dividing by an arbitrary constant the number whose logarithm is expressed.*

121. We may derive the integral (a) directly from Theorem I., thus: In the general form

$$\int x^n dx = \frac{x^{n+1}}{n+1} + h$$

let us determine the constant h by the condition that the integral shall vanish when x has some determinate value a. This gives

$$\frac{a^{n+1}}{n+1} + h = 0; \quad \therefore\ h = -\frac{a^{n+1}}{n+1}.$$

Thus the integral will become

$$\int x^n dx = \frac{x^{n+1} - a^{n+1}}{n+1},$$

in which a takes the place of the arbitrary constant. This expression becomes indeterminate for $n = -1$. But in this case its limit is found by § 71, Ex. 5, to be $\log x - \log a$. Thus we have

$$\int x^{-1}dx = \log x - \log a = \log \frac{x}{a},$$

as before, $\log a$ being now the arbitrary constant.

122. *Exponential Functions.* Since we have

$$d(a^x) = \log a \cdot a^x dx,$$

it follows that we have

$$\int \log a \cdot a^x dx = a^x + h,$$

or, applying Th. II., § 119, to the first member and then dividing by $\log a$,

$$\int a^x dx = \frac{a^x + h}{\log a},$$

which we may write in the form

$$\int a^x dx = \frac{a^x}{\log a} + h,$$

because $\dfrac{h}{\log a}$ is itself a constant which we may represent by h.

123. *The Elementary Forms of Integration.* There is no general method for finding the integral of a given differential. What we have to do, when possible, is to reduce the differential to some form in which we can recognize it as the differential of a known function. For this purpose the following elementary forms, derived by differentiation, should be well memorized by the student. We first write the principal known differentials, and to the left give the integral, found by reversing the process. For perspicuity we repeat the forms already found, and we omit the constants of integration.

$$\therefore d(y^{n+1}) = (n+1)y^n dy, \quad \therefore \int y^n dy = \frac{y^{n+1}}{n+1}. \quad (1)$$

$$\therefore d \cdot \log y = \frac{dy}{y}, \quad \therefore \int \frac{dy}{y} = \log y. \quad (2)$$

$$\therefore d \cdot \sin y = \cos y\, dy, \quad \therefore \int \cos y\, dy = \sin y. \quad (3)$$

$$\therefore d \cdot \cos y = -\sin y\, dy, \quad \therefore \int \sin y\, dy = -\cos y. \quad (4)$$

$$\therefore d \cdot \tan y = \sec^2 y\, dy, \quad \therefore \int \frac{dy}{\cos^2 y} = \tan y. \quad (5)$$

$$\therefore d \cdot \cot y = -\frac{dy}{\sin^2 y}, \quad \therefore \int \frac{dy}{\sin^2 y} = -\cot y. \quad (6)$$

$$\therefore d \cdot \sec y = \frac{\tan y}{\cos y}dy, \quad \therefore \int \frac{\tan y\, dy}{\cos y} = \sec y. \quad (7)$$

$$\therefore d \cdot \sin^{(-1)} y = \frac{dy}{\sqrt{1-y^2}}, \quad \therefore \int \frac{dy}{\sqrt{1-y^2}} = \sin^{(-1)} y. \quad (8)$$

$$\therefore d \cdot \cos^{(-1)} y = -\frac{dy}{\sqrt{1-y^2}}, \quad \therefore \int \frac{-dy}{\sqrt{1-y^2}} = \cos^{(-1)} y. \quad (9)$$

$$\therefore d \cdot \tan^{(-1)} y = \frac{dy}{1+y^2}, \quad \therefore \int \frac{dy}{1+y^2} = \tan^{(-1)} y. \quad (10)$$

$$\therefore d \cdot a^y = a^y \log a\, dy, \quad \therefore \int a^y dy = \frac{a^y}{\log a}. \quad (11)$$

$$\therefore d \cdot \sinh^{(-1)} y = \frac{dy}{\sqrt{y^2+1}},$$

$$\therefore \int \frac{dy}{\sqrt{y^2+1}} = \sinh^{(-1)} y = \log(y + \sqrt{y^2+1}). \quad (12)$$

$$\therefore d \cdot \cosh^{(-1)} y = \frac{dy}{\sqrt{y^2-1}},$$

$$\therefore \int \frac{dy}{\sqrt{y^2-1}} = \cosh^{(-1)} y = \log(y + \sqrt{y^2-1}). \quad (13)$$

$$\therefore d \cdot \tanh^{(-1)} y = \frac{dy}{1-y^2},$$

$$\therefore \int \frac{dy}{1-y^2} = \tanh^{(-1)} y = \frac{1}{2} \log \frac{1+y}{1-y}. \quad (14)$$

CHAPTER II.

INTEGRALS IMMEDIATELY REDUCIBLE TO THE ELEMENTARY FORMS.

124. *Integrals Reducible to the Form* $\int y^n dy$. The following are examples of how, by suitable transformations, we may reduce integrals to the form (1). Let it be required to find

$$\int (a+x)^n dx.$$

We might develop $(a+x)^n$ by the binomial theorem, and then integrate each term separately by applying Theorem III., § 119. But the following is a simpler way. Since we have $dx = d(a+x)$, we may write the integral thus:

$$\int (a+x)^n d(a+x).$$

It is now in the form (1), y being replaced by $a+x$. Hence

$$\int (a+x)^n dx = \frac{(a+x)^{n+1}}{n+1} + h. \qquad (1)$$

In the same way,

$$\int (a-x)^n dx = -\int (a-x)^n d(a-x) = h - \frac{(a-x)^{n+1}}{n+1}.$$

To take another step, let us have to find

$$\int (a+bx)^n dx.$$

We have

$$dx = \frac{1}{b} d(bx) = \frac{1}{b} d(a+bx).$$

Hence, by applying Th. II.,

$$\int (a+bx)^n dx = \frac{1}{b}\int (a+bx)^n d(a+bx) = \frac{(a+bx)^{n+1}}{b(n+1)} + h. \qquad (2)$$

210   THE INTEGRAL CALCULUS.

We might also introduce a new symbol, $y \equiv a + bx$, and then we should have to integrate $y^n dy$ with the result in § 123. Substituting for y its value in terms of x, we should then have the result (2).*

These transformations apply equally whether n, a and b are entire or fractional, positive or negative.

EXERCISES.

Find:

1. $\int (a+x)^4 dx.$ 2. $\int 3(a-x)^3 dx.$

3. $\int (a-2x)^4 dx.$ 4. $\int (a+x)^{-2} dx.$ 5. $\int (a-x)^{-3} dx.$

6. $\int (a+mx)^{-3} dx.$ 7. $\int (a-mx)^3 dx.$ 8. $\int (a-mx)^{-3} dx.$

9. $\int \dfrac{dx}{(a+x)^3}.$ 10. $\int \dfrac{dx}{(a-x)^3}.$ 11. $\int \dfrac{dx}{(a-4x)^3}.$

12. $\int (a+x)^p dx.$ 13. $\int (a+nx)^p dx.$ 14. $\int (a+x^2)^p x\, dx.$

15. $\int \left(\dfrac{a}{x^2} + \dfrac{b}{x^3} + \dfrac{c}{x^4}\right) dx.$ 16. $\int \dfrac{dx}{(a-x)^n}.$

17. $\int \left(\dfrac{b}{(a-x)^2} + \dfrac{c}{(a-x)^3} + \dfrac{e}{(a-x)^4}\right) dx.$

18. $\int \left(\dfrac{b}{(a-mx)^2} + \dfrac{c}{(a-mx)^3} + \dfrac{e}{(a-mx)^4}\right) dx.$

19. $\int (a+bx+cx^2)(b+2cx) dx.$

20. $\int (a+bx+cx^2)^n (b+2cx) dx.$

21. $\int \dfrac{(b+2cx) dx}{(a+bx+cx^2)^n}.$

---

* The question whether to introduce a new symbol for a function whose differential is to be used must be decided by the student in each case. He is advised, as a rule, to first use the function, because he then gets a clearer view of the nature of the transformation. He can then replace the function by a new symbol whenever the labor of repeatedly writing the function will thereby be saved.

125. *Application to the Case of a Falling Body.* We have shown (§ 33) that if, at a time t, a body is at a distance z from a point, the velocity of motion of the body is equal to the derivative $\dfrac{dz}{dt}$. Now, when a body falls from a height under the influence of a uniform force g of gravity, unmodified by any resistance, the law in question asserts that equal velocities are added in equal times. That is, if z be the height of the body above the surface of the earth, and if we count the time t from the moment at which the body began to fall, the law asserts that

$$\frac{dz}{dt} = -gt, \qquad (a)$$

the negative sign indicating that the force g acts so as to diminish the height z.

By integrating this expression, we have

$$z = h - \tfrac{1}{2}gt^2. \qquad (b)$$

Here the constant h represents the height z of the body at the moment when $t = 0$, or when the body began to fall.

From the definition of h and z, it follows that $h - z$ is the distance through which the body has fallen. The equation (b) gives

$$h - z = \tfrac{1}{2}gt^2. \qquad (c)$$

Hence: *The distance through which the body has fallen is proportional to the square of the time.*

At the end of the time t the velocity of the body, measured downwards, is, by (a), equal to gt. If at this moment the velocity became constant, the body would, in another equal interval t, move through the space $gt \times t = gt^2$. Hence, by comparing with (c) we reach by another method a result of § 33, namely:

In any period of time a body falls from a state of rest through half the distance through which it would move in the same period with its acquired velocity at the end of the period.

126. Reduction to the Logarithmic Form. Let us have to find

$$u = \int \frac{m\,dx}{ax+b}.$$

Since $\quad dx = \frac{1}{a}d(ax) = \frac{1}{a}d(ax+b),$

we may write this expression in the form

$$\int \frac{m}{a}\frac{d(ax+b)}{ax+b},$$

and the integral becomes

$$u = \frac{m}{a}\int \frac{d(ax+b)}{ax+b} = \frac{m}{a}\log\frac{ax+b}{c},$$

c being an arbitrary constant.

EXERCISES.

Integrate the following expressions multiplied by dx:

1. $x + \dfrac{1}{x}.$
2. $\dfrac{b}{x}.$
3. $\dfrac{m}{x}.$

4. $\dfrac{1}{x+1}.$
5. $\dfrac{1}{2x-1}.$
6. $\dfrac{m}{cx-b}.$

7. $\dfrac{m}{nx} + x^n.$
8. $\dfrac{a}{2ax+b}.$
9. $\dfrac{a^2}{2bx+a^2}.$

10. $\dfrac{x^2+kx}{4+k}.$
11. $\dfrac{a+b}{ax+b}.$
12. $\dfrac{m-n}{mx-n}.$

13. Find $\int \dfrac{x\,dx}{1+x^2}.$
14. $\int \dfrac{x\,dx}{1+mx^2}.$

Note that $x\,dx = \tfrac{1}{2}d(x^2) = \tfrac{1}{2}d(1+x^2).$

15. $\int \dfrac{x\,dx}{a^2-mx^2}.$
16. $\int \dfrac{x^2\,dx}{1+x^3}.$
17. $\int \dfrac{\log x\,dx}{x}.$

Note that $\log x\,\dfrac{dx}{x} = \log x\,d.\log x.$

18. $\int \dfrac{\log(1+y)}{1+y}dy.$
19. $\int \dfrac{x\,dx}{\sqrt{a^2+x^2}}.$
20. $\int \dfrac{x\,dx}{(1-x^2)^{\frac{1}{2}}}.$

INTEGRALS REDUCIBLE TO ELEMENTARY FORMS. 213

127. *Trigonometric Forms.* The following are examples of the reduction of certain trigonometric forms:

$$\int \cos mx\, dx = \frac{1}{m}\int \cos mx\, d(mx) = \frac{1}{m}\sin mx + h.$$

$$\int \sin mx\, dx = \frac{1}{m}\int \sin mx\, d(mx) = h - \frac{1}{m}\cos mx.$$

$$\int \cos(a+mx)dx = \frac{1}{m}\int \cos(a+mx)d(a+mx)$$
$$= \frac{\sin(a+mx)}{m} + h.$$

$$\int \tan x\, dx = \int \frac{\sin x\, dx}{\cos x} = -\int \frac{d\cdot \cos x}{\cos x}$$
$$= h - \log \cos x = \log c \sec x,$$

where $h = \log c$.

In the same way,

$$\int \cot x\, dx = \log c \sin x.$$

$$\int \frac{dx}{\sin x \cos x} = \int \frac{1}{\tan x}\frac{dx}{\cos^2 x} = \int \frac{d\cdot \tan x}{\tan x} = \log c \tan x.$$

$$\int \frac{dx}{\sin x} = \frac{1}{2}\int \frac{dx}{\sin \tfrac{1}{2}x \cos \tfrac{1}{2}x} = \log c \tan \frac{1}{2}x.$$

$$\int \frac{dx}{\cos x} = \int \frac{dx}{\sin(\tfrac{1}{2}\pi - x)} = \log c \tan\left(\frac{\pi}{4} - \frac{x}{2}\right).$$

EXERCISES.

Integrate:

1. $(1 + \cos y)dy.$
2. $(1 - e \sin u)du.$
3. $\cos 2y\, dy.$ Ans. $\tfrac{1}{2}\int \cos 2y\, d(2y) = \tfrac{1}{2}\sin 2y.$
4. $\sin 2y\, dy.$
5. $\cos ny\, dy.$
6. $\sin y \cos y\, dy.$ Ans. $\tfrac{1}{4}\int \sin 2y\, d(2y) = -\tfrac{1}{4}\cos 2y.$
7. $\tan 2x\, dx.$
8. $\cot 2x\, dx.$
9. $2\cos^2 x\, dx.$

   Ans. $\int (1 + \cos 2x)dx = x + h + \tfrac{1}{2}\sin 2x.$

10. $2\sin^2 x\, dx.$
11. $\tan 2y\, dy.$
12. $\dfrac{\cos y\, dy}{1+\sin y}.$ *Ans.* $\displaystyle\int\dfrac{d(1+\sin y)}{1+\sin y}=\log c(1+\sin y).$
13. $\dfrac{\sin y\, dy}{1+\cos y}.$
14. $\dfrac{\sin y\, dy}{1-\cos y}.$
15. $\dfrac{\cos y\, dy}{1-\sin y}.$
16. $\dfrac{\sin 2y\, dy}{\cos y}.$
17. $\dfrac{\cos^2 x-\sin^2 x}{\sin 2x}\, dx.$
18. $\dfrac{\sin 2x}{\cos^2 x-\sin^2 x}\, dx.$
19. $\dfrac{dx}{\cos mx}.$
20. $\dfrac{dx}{\sin mx}.$
21. $\dfrac{dx}{\sin mx \cos mx}.$
22. $\sin(mx+a)dx.$
23. $\cos(a-nx)dx.$
24. $\tan nx\, dx.$
25. $\tan(2x-a)dx.$
26. $\dfrac{dx}{\sin(a-x)}.$
27. $\dfrac{dx}{\cos(b-nx)}.$
28. $\dfrac{dx}{\sin(a-nx)}.$
29. $\dfrac{\cos ny\, dy}{a+\sin ny}.$
30. $\dfrac{\sin ny\, dy}{a-\cos ny}.$
31. $\dfrac{\sec^2 x\, dx}{a-m\tan x}.$

128. *Integration of* $\dfrac{dx}{a^2+x^2}$ *and* $\dfrac{dx}{a^2-x^2}.$

We see at once that the first differential may be reduced to that of an inverse tangent; thus,

$$\frac{dx}{x^2+a^2}=\frac{1}{a^2}\frac{dx}{\frac{x^2}{a^2}+1}=\frac{1}{a}\frac{d\left(\frac{x}{a}\right)}{\frac{x^2}{a^2}+1}.$$

Hence

$$\int\frac{dx}{a^2+x^2}=\frac{1}{a}\int\frac{d\frac{x}{a}}{\frac{x^2}{a^2}+1}=\frac{1}{a}\tan^{(-1)}\frac{x}{a}+h. \qquad (1)$$

We find in the same way

$$\int\frac{dx}{a^2-x^2}=\frac{1}{a}\tan h^{(-1)}\frac{x}{a}+h=\frac{1}{2a}\log c\frac{a+x}{a-x}, \qquad (2)$$

c being an arbitrary constant factor.

INTEGRALS REDUCIBLE TO ELEMENTARY FORMS. 215

129. *Integrals of the form* $\int \dfrac{dx}{a + bx + cx^2}$.

The reduction of integrals of this form depends upon the character of the roots of the quadratic equation

$$cx^2 + bx + a = 0. \qquad (1)$$

I. If these roots are imaginary, the integral is the inverse of a trigonometric tangent.

II. If the roots are real and unequal, the integral is the inverse of an hyperbolic tangent.

III. If the roots are real and equal, that is, if the above expression is a perfect square, the integral is an algebraic function.

Dividing the denominator of the fraction by the coefficient of x^2, the given integral may be written

$$\frac{1}{c} \int \frac{dx}{x^2 + \dfrac{bx}{c} + \dfrac{a}{c}}. \qquad (a)$$

Writing $2p$ for $\dfrac{b}{c}$ and q for $\dfrac{a}{c}$, the expression to be integrated may be reduced to one of the forms of § 128, thus:

$$\frac{dx}{x^2 + 2px + q} = \frac{dx}{(x+p)^2 + q - p^2} = \frac{d(x+p)}{(x+p)^2 + q - p^2}. \qquad (b)$$

The three cases now depend on the sign of $q - p^2$.

I. If $q - p^2$ is positive, the roots of (1) are imaginary and the form is the first of the last article with $x + p$ in the place of x, and $q - p^2$ in the place of a^2. Hence we have

$$\int \frac{dx}{x^2 + 2px + q} = \int \frac{d(x+p)}{(x+p)^2 + q - p^2}$$
$$= \frac{1}{\sqrt{q - p^2}} \tan^{(-1)} \frac{x + p}{\sqrt{q - p^2}} + h. \qquad (1)$$

Comparing this result with (a), we see that this integral may be reduced to its primitive form by changing p into

$\frac{1}{2}\frac{b}{c}$ and q into $\frac{a}{c}$. Substituting and reducing, we have

$$\int \frac{dx}{a+bx+cx^2} = \frac{1}{c}\int \frac{dx}{x^2+\frac{b}{c}x+\frac{a}{c}}$$

$$= \frac{1}{c}\cdot\frac{1}{\sqrt{\frac{a}{c}-\frac{b^2}{4c^2}}}\tan^{(-1)}\frac{x+\frac{b}{2c}}{\sqrt{\frac{a}{c}-\frac{b^2}{4c^2}}}$$

$$= \frac{2}{\sqrt{4ac-b^2}}\tan^{(-1)}\frac{2cx+b}{\sqrt{4ac-b^2}} + h. \quad (2)$$

II. If $q - p^2$ is negative, that is, if $4ac - b^2$ is negative in (2), the expression (2) will contain two imaginary quantities. But these two quantities cancel each other, so that the expression is always real. When $q - p^2$ is negative, we write (b) in the form

$$-\frac{d(x+p)}{p^2 - q - (x+p)^2}.$$

The integral is now in the form (2) of § 128, and we have

$$\int \frac{dx}{x^2 + 2px + q} = -\int \frac{d(x+p)}{p^2 - q - (x+p)^2}$$

$$= h - \frac{1}{\sqrt{p^2-q}}\tanh^{(-1)}\frac{x+p}{\sqrt{p^2-q}}$$

$$= h - \frac{1}{2\sqrt{p^2-q}}\log c\frac{\sqrt{p^2-q}+x+p}{\sqrt{p^2-q}-(x+p)}. \quad (3)$$

Making the same substitutions in these equations that we made in Case I., we find

$$\int \frac{dx}{a+bx+cx^2} = h - \frac{2}{\sqrt{b^2-4ac}}\tan h^{(-1)}\frac{2cx+b}{\sqrt{b^2-4ac}}$$

$$= h - \frac{1}{\sqrt{b^2-4ac}}\log c\frac{\sqrt{b^2-4ac}+2cx+b}{\sqrt{b^2-4ac}-(2cx+b)}. (4)$$

III. If $p^2 - q = 0$, the expression to be integrated becomes $\frac{dx}{(x+p)^2}$. We have already integrated this form and found

$$\int \frac{dx}{(x+p)^2} = h - \frac{1}{x+p}.$$

INTEGRALS REDUCIBLE TO ELEMENTARY FORMS. 217

EXERCISES.

Integrate the following expressions:

1. $\dfrac{dx}{x^2 - 2x - 4}$. 2. $\dfrac{dx}{(x-\alpha)(x-\beta)}$. 3. $\dfrac{dx}{a + 2bx - x^2}$.

4. $\dfrac{dx}{x^2 + 4x + 2}$. 5. $\dfrac{dx}{x(x-a)}$. 6. $\dfrac{dx}{(a-x)(x-b)}$.

130. *Inverse Sines and Cosines as Integrals.* From what has already been shown (§ 123, (8) and (9)), it will be seen that we have the two following integral forms:

$$\int \frac{dx}{\sqrt{1-x^2}} = \sin^{(-1)} x + h \equiv u; \qquad (a)$$

$$\int -\frac{dx}{\sqrt{1-x^2}} = \cos^{(-1)} x + h' \equiv u'; \qquad (b)$$

where we have added h and h' as arbitrary constants of integration.

Comparing the first members of these equations, we see that each is the negative of the other. The question may therefore be asked why we should not write the second equation in the form

$$u' = -\int \frac{dx}{\sqrt{1-x^2}} = h'' - \sin^{(-1)} x, \qquad (c)$$

as well as in the form (b). The answer is that no error would arise in doing so, because the forms (b) and (c) are equivalent. From (b) we derive

$$x = \cos(u' - h') = \cos(h' - u'); \qquad (d)$$

and from (c),

$$x = \sin(h'' - u'). \qquad (e)$$

Now, we always have $\sin(a + 90°) = \cos a$. Hence (d) and (e) become identical by putting

$$h'' = h' + 90°,$$

which we may always do, because the value of h'' is quite arbitrary.

131. The preceding reasoning illustrates the fact that integrals expressed by circular functions may be expressed either in the direct or inverse form. That is, if the relation between the differentials of u and x is expressed in the form

$$du = \frac{dx}{\sqrt{1-x^2}},$$

we may express the relation between u and x themselves either in the form

$$u = \sin^{(-1)} x + h$$

or in the form

$$x = \sin(u - h).$$

So, also, in the form (1) of § 128 we may express the relation between x and u either as it is there written or in the reverse form,

$$x = a \tan a(u - h).$$

132. *Integration of* $\dfrac{dx}{\sqrt{a^2 \mp x^2}}$.

We have

$$\int \frac{dx}{\sqrt{a^2 - x^2}} = \int \frac{d \cdot \dfrac{x}{a}}{\sqrt{1 - \dfrac{x^2}{a^2}}} = \sin^{(-1)} \frac{x}{a} + h. \quad (1)$$

In the same way

$$\int \frac{-dx}{\sqrt{a^2 - x^2}} = \cos^{(-1)} \frac{x}{a} + h \quad \text{or} \quad h - \sin^{(-1)} \frac{x}{a}. \quad (2)$$

We also have

$$\int \frac{dx}{\sqrt{a^2 + x^2}} = \int \frac{d \cdot \dfrac{x}{a}}{\sqrt{1 + \dfrac{x^2}{a^2}}} = \sin h^{(-1)} \frac{x}{a} + h$$

$$= \log \frac{c}{a}(x + \sqrt{x^2 + a^2}). \quad (3)$$

INTEGRALS REDUCIBLE TO ELEMENTARY FORMS. 219

$$\int \frac{dx}{\sqrt{x^2-a^2}} = \int \frac{d \cdot \frac{x}{a}}{\sqrt{\frac{x^2}{a^2}-1}} = \cosh^{(-1)} \frac{x}{a} + h$$

$$= \log \frac{c}{a}(x + \sqrt{x^2-a^2}). \quad (4)$$

EXERCISES.

Integrate the differentials:

1. $\dfrac{dx}{\sqrt{c-x^2}}$.

2. $\dfrac{dy}{\sqrt{4a^2-y^2}}$.

3. $\dfrac{n\,dy}{\sqrt{a^2-n^2y^2}}$.

4. $\dfrac{dx}{\sqrt{a^2-(x-a)^2}}$.

5. $\dfrac{m\,dz}{\sqrt{4a^2-m^2z^2}}$.

6. $\dfrac{dz}{\sqrt{4a^2-m^2z^2}}$.

7. $\dfrac{dx}{\sqrt{4c^2+x^2}}$.

8. $\dfrac{m\,dx}{\sqrt{a^2+m^2x^2}}$.

9. $\dfrac{dy}{\sqrt{4a^2+9y^2}}$.

10. $\dfrac{dx}{\sqrt{a^2+m^2(x-a)^2}}$.

11. $\dfrac{dx}{\sqrt{b+c(x-a)^2}}$.

12. $\dfrac{dx}{\sqrt{(x-a)^2-4c^2}}$.

13. $\dfrac{2x\,dx}{\sqrt{a^4-x^4}}$.

14. $\dfrac{nx^{n-1}\,dx}{\sqrt{a^{2n}-x^{2n}}}$.

15. If $du = \dfrac{-\cos x\,dx}{\sqrt{a^2-\sin^2 x}}$, then $\sin x = a \cos(u+h)$.

16. $\dfrac{e^x\,dx}{\sqrt{1-e^{2x}}}$.

17. $\dfrac{dx}{x\sqrt{1-(\log x)^2}}$.

18. $\dfrac{-\sin x\,dx}{a^2+\cos^2 x}$.

19. $\dfrac{\cos x\,dx}{a^2+\sin^2 x}$.

20. $\dfrac{(x-a)\,dx}{\sqrt{1-(x-a)^4}}$.

21. $\dfrac{(x+a)\,dx}{\sqrt{1+(x+a)^4}}$.

133. *Integration of* $\dfrac{dx}{\sqrt{a + bx \pm cx^2}}$. Every differential of this form can be reduced to one of the three forms of the preceding article by a process similar to that of §129. The mode of reduction will depend upon the sign of the term cx^2.

CASE I. *The term cx^2 is negative.* Putting, as before,

$$p \equiv \frac{1}{2}\frac{b}{c}, \quad q \equiv \frac{a}{c},$$

we have

$$\sqrt{a + bx - cx^2} = \sqrt{c}\,\sqrt{q + 2px - x^2} = \sqrt{c}\,\sqrt{p^2 + q - (x-p)^2}.$$

Then, comparing with (1) of §132, we find

$$\int \frac{dx}{\sqrt{a + bx - cx^2}} = \frac{1}{\sqrt{c}} \int \frac{d(x - p)}{\sqrt{p^2 + q - (x - p)^2}}$$

$$= \frac{1}{\sqrt{c}} \sin^{(-1)} \frac{x - p}{\sqrt{p^2 + q}} = \frac{1}{\sqrt{c}} \sin^{(-1)} \frac{2cx - b}{\sqrt{b^2 + 4ac}}. \quad (1)$$

In order that this expression may be real, $p^2 + q$ or $b^2 + 4ac$ must be positive. If this quantity is negative the integral will be wholly imaginary, but may be reduced to an inverse hyperbolic sine multiplied by the imaginary unit.

CASE II. *The term cx^2 is positive.* We now have

$$\sqrt{a + bx + cx^2} = \sqrt{c}\,\sqrt{(x+p)^2 + q - p^2}.$$

$$\int \frac{dx}{\sqrt{a + bx + cx^2}} = \frac{1}{\sqrt{c}} \int \frac{d(x + p)}{\sqrt{(x+p)^2 + q - p^2}}$$

$$= \frac{1}{\sqrt{c}} \log C(x + p + \sqrt{x^2 + 2px + q})$$

$$= \frac{1}{c^{\frac{1}{2}}} \log \frac{C}{2c^{\frac{1}{2}}}(2cx + b + 2c^{\frac{1}{2}}\sqrt{a + bx + cx^2}).$$

Because C is an arbitrary quantity, the quotient of C by $2c^{\frac{1}{2}}$ is equally an arbitrary quantity, and may be represented by the single symbol C. Thus we have

$$\int \frac{dx}{\sqrt{a + bx + cx^2}} = \frac{1}{c^{\frac{1}{2}}} \log C(b + 2cx + 2\sqrt{c}\,\sqrt{a + bx + cx^2}). \quad (2)$$

INTEGRALS REDUCIBLE TO ELEMENTARY FORMS. 221

EXERCISES.

Integrate:

1. $\dfrac{dy}{\sqrt{4a^2+4by-y^2}}.$

2. $\dfrac{dy}{\sqrt{(a+y)(b-y)}}.$

3. $\dfrac{y\,dy}{\sqrt{8-4y^2+y^4}}.$

4. $\dfrac{dy}{\sqrt{a^2y^2-by+b^2}}.$

5. $\dfrac{\cos\theta\,d\theta}{\sqrt{1-\sin\theta-\sin^2\theta}}.$

6. $\dfrac{\cos\theta\,d\theta}{\sqrt{1-\sin\theta+\cos^2\theta}}.$

7. $\dfrac{\sin\theta\cos\theta\,d\theta}{\sqrt{4-\cos 2\theta-\cos^2 2\theta}}.$

8. $\dfrac{a\sin\theta\,d\theta}{\sqrt{a^2-b^2(1-\cos\theta)^2}}.$

134. Exponential Forms. Using the form (11) of § 123, we may reduce and integrate the simplest exponential differentials as follows:

$$\int a^{mx}\,dx = \frac{1}{m}\int a^{mx}\,d(mx) = \frac{a^{mx}}{m\log a} + h. \quad (1)$$

$$\int a^{x+b}\,dx = \int a^{x+b}\,d(x+b) = \frac{a^{x+b}}{\log a} + h. \quad (2)$$

$$\int a^{mx+b}\,dx = \frac{1}{m}\int a^{mx+b}\,d(mx+b) = \frac{a^{mx+b}}{m\log a} + h. \quad (3)$$

$$\int a^{-mx}\,dx = -\frac{1}{m}\int a^{-mx}\,d(-mx) = \frac{h-a^{-mx}}{m\log a}. \quad (4)$$

EXERCISES.

Integrate:

1. $e^x\,dx.$
2. $b^y\,dy.$
3. $a^{y-1}\,dy.$
4. $(a+b)e^x\,dx.$
5. $a^{y-c}\,dy.$
6. $a^{-x}\,dx.$
7. $(a^x+a^{-x})\,dx.$
8. $(a^x-a^{-x})\,dx.$
9. $(a+e^x)\,dx.$
10. $(a^{kx}-a^{-kx})\,dx.$
11. $\dfrac{e^x\,dx}{1+e^x}.$
12. $\dfrac{e^{mx}\,dx}{1+e^{mx}}.$
13. $\dfrac{(e^x-e^{-x})\,dx}{e^x+e^{-x}}.$
14. $(1+a^x)^2\,dx.$
15. $(a^{mx}+a^{-mx})^2\,dx.$
16. $\int e^{x^2}x\,dx.$

CHAPTER III.

INTEGRATION BY RATIONAL TRANSFORMATIONS.

135. We have now to consider certain forms which cannot be reduced so simply and directly as those treated in the last chapter. Before passing to general methods we shall consider some simple cases.

I. *Integration of* $\dfrac{(a+x)^m}{x^n}dx$. Any form of this kind, when m is entire, may be integrated by developing the numerator by the binomial theorem. We then have

$$\frac{(a+x)^m}{x^n} = \frac{a^m}{x^n} + \frac{ma^{m-1}x}{x^n} + \binom{m}{2}\frac{a^{m-2}x^2}{x^n} + \text{etc.},$$

and each term can be integrated separately. If $n < m+2$, and entire, one of the terms of the integral will contain $\log x$.

II. *Integration of* $\dfrac{x^m dx}{(a+bx)^n}$. We may reduce this form to the preceding, by introducing a new variable, z, defined by the equation

$$z \equiv a + bx.$$

This gives $\quad x = \dfrac{z-a}{b}; \quad dx = \dfrac{dz}{b}.$

Substituting these values of x and dx in the expression to be integrated, it becomes

$$\frac{(z-a)^m dz}{b^{m+1}z^n},$$

which may be integrated by the method of the last article.

III. *Integration of* $\dfrac{x dx}{a+bx \pm cx^2}$. We reduce the denomi-

INTEGRATION BY RATIONAL TRANSFORMATIONS.

nator to the form $\pm (y^2 - q) \pm (x + p)^2$ as in §129. Then, putting, for brevity,

$$b^2 \equiv p^2 - q,$$
$$z \equiv x + p,$$

which gives $\quad dx = dz,$

the integration will have to be performed on an expression of the form

$$\frac{(z-p)dx}{b^2 \pm z^2} = \frac{zdz}{b^2 \pm z^2} - \frac{pdz}{b^2 \pm z^2}.$$

Each of these terms may be integrated by methods already given (§§ 126, 128).

The process is exactly the same if we have to find

$$\int \frac{(a+bx)dx}{b^2 \pm (x-p)^2}.$$

EXERCISES.

Integrate:

1. $\dfrac{(x-a)^2 dx}{x^2}.$

2. $\dfrac{\left(\dfrac{1}{a} - \dfrac{1}{x}\right)^2 dx}{x}.$

3. $\dfrac{x^2 dx}{(a-x)^4}.$

4. $\dfrac{x^2 dx}{(1+x^2)^4}.$

5. $\dfrac{dx}{x^2\left(\dfrac{1}{a}+\dfrac{1}{x}\right)^3}.$

6. $\dfrac{(x+a)dx}{(a-x)^2}.$

7. $\dfrac{x^5 dx}{(a^2-x^2)^2}.$

8. $\dfrac{x^2 dx}{\left(\dfrac{1}{a^2} - \dfrac{1}{x^2}\right)^2}.$

9. $\dfrac{xdx}{a^2+(b-x)^2}.$

10. $\dfrac{zdz}{(a+z)^2+(a-z)^2}.$

11. $\dfrac{(y-b)dy}{(y-b)^2+(y+b)^2}.$

12. $\dfrac{(z-c)dz}{a^2-az+z^2}.$

13. $\dfrac{(x-a)dx}{x(x-b)}.$

14. $\dfrac{(y+a)dy}{a^2-(y+b)^2}.$

15. $\dfrac{z^n dz}{(1+z)^n}.$

16. $\dfrac{z^2 dz}{(1-z)^4}.$

136. *Reduction of Rational Fractions in general.* A rational fraction is a fraction whose numerator and denominator are entire functions of the variable. The general form is

$$\frac{p_0 + p_1 x + p_2 x^2 + \ldots + p_m x^m}{q_0 + q_1 x + q_2 x^2 + \ldots + q_n x^n} = \frac{N}{D}.$$

If the degree m of the numerator exceeds the degree n of the denominator, we may divide the numerator by the denominator until we have a remainder of less degree than n. Then, if we put Q for the entire part of the quotient, and R for the remainder, the fraction will be reduced to

$$\frac{N}{D} = Q + \frac{R}{D}.$$

If we have to integrate this expression, then, since Q is an entire function of x, the differential Qdx can be integrated by § 119, leaving only the proper fraction $\frac{R}{D}$. Now, such a fraction always admits of being divided into the sum of a series of partial fractions with constant numerators, provided that we can find the roots of the equation $D = 0$. The theory of this process belongs to Algebra, but we shall show by examples how to execute it in the three principal cases which may arise.

CASE I. *The roots of the equation $D = 0$ all real and unequal.* Let these roots be $\alpha, \beta, \gamma \ldots \theta$. Then, as shown in Algebra, we shall have

$$D = (x - \alpha)(x - \beta)(x - \gamma) \ldots (x - \theta).$$

We then assume

$$\frac{R}{D} = \frac{A}{x - \alpha} + \frac{B}{x - \beta} + \frac{C}{x - \gamma} + \ldots,$$

A, B, C, etc., being undetermined coefficients. To determine them we reduce the fractions in the second member to the common denominator D, equate the sum of the numerators of the new fractions to R, and then equate the coefficients of like powers of x.

INTEGRATION BY RATIONAL TRANSFORMATIONS.

As an example, let us take the fraction
$$\frac{x+3}{x^3-x}dx.$$
We readily find, by solving the equation $x^3 - x = 0$,
$$x^3 - x = x(x-1)(x+1).$$
Assume
$$\frac{x+3}{x^3-x} = \frac{A}{x} + \frac{B}{x-1} + \frac{C}{x+1}$$
$$= \frac{(A+B+C)x^2 + (B-C)x - A}{x^3 - x}.$$
Equating the coefficients of powers of x, we have
$$A + B + C = 0;$$
$$B - C = 1;$$
$$A = -3;$$
whence $B = 2$ and $C = 1$. Hence
$$\frac{x+3}{x^3-x} = -\frac{3}{x} + \frac{2}{x-1} + \frac{1}{x+1};$$
and then, by § 120,
$$\int\frac{x+3}{x^3-x}dx = -3\int\frac{dx}{x} + 2\int\frac{dx}{x-1} + \int\frac{dx}{x+1}$$
$$= -3\log x + 2\log(x-1) + \log(x+1) + \log C$$
$$= \log\frac{C(x+1)(x-1)^2}{x^3}.$$

EXERCISES.

Integrate:

1. $\dfrac{(x-1)dx}{x^3-x-6}.$

2. $\dfrac{xdx}{x^4-1}.$

3. $\dfrac{xdx}{1-x^4}.$

4. $\dfrac{(x+x^3)dx}{(x-1)(x+1)(x-2)(x+2)}.$

5. $\dfrac{x^2+x+1}{x^3+x^2-6x}.$

6. $\dfrac{x^2 dx}{x^3-a^3}.$

7. $\dfrac{(x^4+2x^3)dx}{x^3+2x^2-8x}.$

8. $\dfrac{(x^4+x^2)dx}{x(x-1)(x+1)(x-2)}.$

$x^3 dx$

dx

CASE II. *Some of the roots equal to each other.* Let the factor $x - \alpha$ appear in D to the nth power. Then, if we followed the process of Case I., we should find ourselves with more equations than unknown quantities, because the n fractions

$$\frac{A}{x-\alpha} + \frac{B}{x-\alpha} + \frac{C}{x-\alpha} + \cdots$$

would coalesce into one. To avoid this we write the assumed series of fractions in the form

$$\frac{A}{(x-\alpha)^n} + \frac{B}{(x-\alpha)^{n-1}} + \cdots + \frac{F}{x-\alpha} + \frac{H}{x-\beta} + \text{etc.},$$

and then we proceed to reduce to a common denominator as before. The coefficients A, B, etc., are now equal in number to the terms of the equation $D = 0$, so that we shall have exactly conditions enough to determine them.

As an example, let it be required to integrate

$$\frac{x^2 - 5}{x^3 - x^2 - x + 1} dx.$$

We have $x^3 - x^2 - x + 1 = (x-1)^2(x+1)$.

We then assume

$$\frac{x^2-5}{(x-1)^2(x+1)} = \frac{A}{(x-1)^2} + \frac{B}{x-1} + \frac{C}{x+1}$$
$$= \frac{(B+C)x^2 + (A-2C)x + A - B + C}{(x-1)^2(x+1)}.$$

We find, by equating and solving,

$$A = -2;$$
$$B = +2;$$
$$C = -1.$$

Hence

The required integral is

$$-2\int (x-1)^{-2}dx + 2\int \frac{dx}{x-1} - \int \frac{dx}{x+1}$$

$$= \frac{2}{x-1} + 2\log(x-1) - \log(x+1) + \log C$$

$$= \frac{2}{x-1} + \log \frac{C(x-1)^2}{x+1}.$$

EXERCISES.

Integrate:

1. $\dfrac{dx}{x(x+1)^2}.$

2. $\dfrac{dx}{x^2(x-1)^2}.$

3. $\dfrac{x^2 dx}{(x-1)^2(x+2)^2}.$

4. $\dfrac{dx}{(x-a)^2(x-b)^2}.$

5. $\dfrac{(a+x)dx}{x^2(x-a)^2}.$

6. $\dfrac{(a-x)dx}{x^2(x+a)^2(x-b)}.$

CASE III. *Imaginary roots.* Were the preceding methods applied without change to the case when the equation $D = 0$ has imaginary roots, we should have a result in an imaginary form, though actually the integral is real. We therefore modify the process as follows:

It is shown in Algebra that imaginary roots enter an equation in pairs, so that if $x = \alpha + \beta i$ (where $i \equiv \sqrt{-1}$) is a root, then $x = \alpha - \beta i$ will be another root. To these roots correspond the product

$$(x - \alpha - \beta i)(x - \alpha + \beta i) = (x - \alpha)^2 + \beta^2.$$

By thus combining the imaginary factors the function D will be divided into factors all of which are real, but some of which, in the case of imaginary roots, will be of the second degree.

The assumed fraction corresponding to a pair of imaginary roots we place in the form

$$\frac{A + Bx}{(x-\alpha)^2 + \beta^2},$$

and then proceed to determine A and B as before by equations of condition. We then divide the numerator $A + Bx$ into the two parts

$$A + B\alpha \quad \text{and} \quad B(x - \alpha),$$

the sum of which is $A + Bx$. Thus we have to integrate

$$\int \frac{A + B\alpha}{(x - \alpha)^2 + \beta^2} dx + \int \frac{B(x - \alpha)dx}{(x - \alpha)^2 + \beta^2}. \quad (a)$$

The first term of (a) is, by methods already developed,

$$\frac{A + B\alpha}{\beta} \tan^{(-1)} \frac{x - \alpha}{\beta},$$

and the second is

$$\tfrac{1}{2} B \log \left((x - \alpha)^2 + \beta^2 \right).$$

We therefore have, for the complete integral,

$$\int \frac{A + Bx}{(x - \alpha)^2 + \beta^2} = \frac{A + B\alpha}{\beta} \tan^{(-1)} \frac{x - \alpha}{\beta}$$
$$+ \tfrac{1}{2} B \log\{(x - \alpha)^2 + \beta^2\} + h.$$

EXERCISES.

1. $\int \dfrac{x + 3x^2}{x^4 - 1} dx.$
2. $\int \dfrac{dx}{x^3 - 1}.$

The real factors of the denominator in Ex. 1 are $(x^2 + 1)(x + 1)(x - 1)$. We resolve the given fraction in the form

$$\frac{A + Bx}{x^2 + 1} + \frac{C}{x + 1} + \frac{D}{x - 1},$$

and find it equal to $\dfrac{x}{x^2 + 1} + \dfrac{1}{x + 1} + \dfrac{1}{x - 1}$. Then the integral is found to be $\tfrac{1}{2} \log (x^2 + 1) + \log (x^2 - 1)$.

The factors of the denominator in Ex. 2 are $x - 1$ and $x^2 + x + 1 = (x + \tfrac{1}{2})^2 + \tfrac{3}{4}$.

3. $\int \dfrac{dx}{x^3 + 1}.$
4. $\int \dfrac{(x^2 + 1)dx}{x^3 - 2x + 4}.$

Note that $x + 2$ is a factor of the denominator in (4).

137. Integration by Parts. Let u and v be any two functions of x. We have

$$\frac{d(uv)}{dx} = u\frac{dv}{dx} + v\frac{du}{dx}.$$

By transposing and integrating we have

$$\int u\frac{dv}{dx}dx = uv - \int v\frac{du}{dx}dx + h, \quad (1)$$

which is a general formula of the widest application, and should be thoroughly memorized by the student. It shows us that whenever the differential function to be integrated can be divided into two factors, one of which $\left(\frac{dv}{dx}dx\right)$ can be integrated by itself, the problem can be reduced to the integration of some new expression $\left(v\frac{du}{dx}dx\right)$.

The formula may be written and memorized in the simpler form

$$\int u\,dv = uv - \int v\,du, \quad (2)$$

it being understood that the expressions dv and du mean differentials with respect to the independent variable, whatever that may be.

It does not follow that the new expression will be any easier to integrate than the original one; and when it is not, the method of integrating by parts will not lead us to the integral. The cases in which it is applicable can only be found by trial.

The general rule embodied in the formulæ (1) and (2) is this:

Express the given differential as the product of one function into the differential of a second function.

Then its integral will be the product of these two functions, minus the integral of the second function into the differential of the first.

EXAMPLES AND EXERCISES IN INTEGRATION BY PARTS.

1. To integrate $x \cos x dx$.

We have $\cos x dx = d \cdot \sin x$.

Therefore in (2) we have

$$u \equiv x; \quad v \equiv \sin x;$$

and the formula becomes

$$\int x \cos x dx = \int x d \cdot \sin x = x \sin x - \int \sin x dx$$
$$= x \sin x + \cos x + h,$$

which is the required expression, as we may readily prove by differentiation.

Show in the same way that—

2. $\int x \sin x dx = - x \cos x + \sin x + h.$

3. $\int x \sec^2 x dx = x \tan x - $ (what?).

4. $\int x \sin x \cos x dx = - \tfrac{1}{4} x \cos 2x + \tfrac{1}{4} \sin 2x + h.$

5. $\int \log x dx = x \log x - \int x d \cdot \log x = x \log x - x + h.$

6. The process in question may be applied any number of times in succession. For example,

$$\int x^2 \cos x dx = \int x^2 d \cdot \sin x = x^2 \sin x - 2 \int x \sin x dx.$$

Then, by integrating the last term by parts, which we have already done,

$$\int x^2 \cos x dx = x^2 \sin x + 2x \cos x - 2 \sin x + h.$$

7. In the same way,

$$\int x^3 \cos x dx = \int x^3 d \cdot \sin x = x^3 \sin x - 3 \int x^2 \sin x dx;$$

$$\int x^2 \sin x dx = -\int x^2 d \cdot \cos x = - x^2 \cos x + 2 \int x \cos x dx.$$

INTEGRATION BY RATIONAL TRANSFORMATIONS. 231

Then, by substitution,

$$\int x^3 \cos x\, dx = (x^3 - 6x) \sin x + (3x^2 - 6) \cos x + h.$$

8. In general,

$$\int x^n \cos x\, dx = \int x^n d \cdot \sin x = x^n \sin x - n \int x^{n-1} \sin x\, dx;$$

$$-\int x^{n-1} \sin x\, dx = \int x^{n-1} d \cdot \cos x$$

$$= x^{n-1} \cos x - (n-1) \int x^{n-2} \cos x\, dx;$$

$$-\int x^{n-2} \cos x\, dx = -x^{n-2} \sin x + (n-2) \int x^{n-3} \sin x\, dx;$$

$$\int x^{n-3} \sin x\, dx = -x^{n-3} \cos x + (n-3) \int x^{n-4} \cos x\, dx.$$

etc.   etc.   etc.

Then, by successive substitution, we find, for the required integral,

$$\{x^n - n(n-1)x^{n-2} + n(n-1)(n-2)(n-3)x^{n-4} - \ldots\} \sin x$$
$$+ \{nx^{n-1} - n(n-1)(n-2)x^{n-3} + \ldots\} \cos x.$$

9. In the same way, show that

$$\int x^n \sin x\, dx =$$
$$\{-x^n + n(n-1)x^{n-2} - n(n-1)(n-2)(n-3)x^{n-4} + \ldots\} \cos x$$
$$+ \{nx^{n-1} - n(n-1)(n-2)x^{n-3} + \ldots\} \sin x.$$

10. $\int x^n \log x\, dx = \dfrac{1}{n+1} \int \log x\, d \cdot (x^{n+1}) = \dfrac{x^{n+1}}{n+1} \log x$

$$- \dfrac{1}{n+1} \int \dfrac{x^{n+1}}{x} dx = \dfrac{x^{n+1}}{n+1} \log x - \dfrac{x^{n+1}}{(n+1)^2}.$$

11. $\int x e^{-ax} dx = -\int \dfrac{1}{a} x\, d \cdot (e^{-ax}) = -\dfrac{x e^{-ax}}{a} + \dfrac{1}{a} \int e^{-ax} dx.$

Now, we have $\int e^{-ax} dx = -\dfrac{e^{-ax}}{a}.$

12. To integrate $x^m e^{-ax} dx$ when m is a positive integer, we proceed in the same way, and repeat the process until we reduce the exponent of x to unity. Thus,

$$\int x^m e^{-ax} dx = -\frac{x^m e^{-ax}}{a} + \frac{m}{a}\int x^{m-1} e^{-ax} dx.$$

Treating this last integral in the same way, and repeating the process, the integral becomes

$$-\frac{x^m e^{-ax}}{a} - \frac{mx^{m-1} e^{-ax}}{a^2} - \frac{m(m-1)x^{m-2} e^{-ax}}{a^3} - \text{etc.}$$

$$= -\frac{e^{-ax}}{a^{m+1}}(a^m x^m + m a^{m-1} x^{m-1} + m(m-1) a^{m-2} x^{m-2} + \ldots + m!).$$

13. From the result of Ex. 5 show that

$$\int (\log x)^2 dx = x(l^2 - 2l + 2) + h,$$

where we put, for brevity, $l \equiv \log x$.

14. Show that, in general, if we put

$$u_n = \int (\log x)^n dx,$$

then
$$u_n = xl^n - nu_{n-1};$$

and therefore, by successive substitution,

$$u_n = x(l^n - nl^{n-1} + n(n-1)l^{n-2} - \ldots \pm n!) + h.$$

15. Deduce $(m+1)\int Px^m dx = Px^{m+1} - \int x^{m+1} dP.$

16. Show that if $\int P dx = Q,$

then
$$\int Px^n dx = Qx^n - n\int Qx^{n-1} dx.$$

Also, if we have

$$\int Q dx = R; \quad \int R dx = S, \text{ etc.,}$$

then

$$\int Px^n dx = Qx^n - nRx^{n-1} + n(n-1)Sx^{n-2} - \text{etc.}$$

CHAPTER IV.

INTEGRATION OF IRRATIONAL ALGEBRAIC DIFFERENTIALS.

138. *When Fractional Powers of the Independent Variable enter into the Expression.* In this case we may render the expression rational by reducing the exponents to their least common denominator, and equating the variable to a new variable raised to the power represented by this denominator.

EXAMPLE. If we have to integrate

$$\frac{1 + x^{\frac{1}{2}}}{1 + x^{\frac{1}{3}}} dx,$$

then, the L. C. D. of the denominators of the exponents being 6, we substitute for x the new variable z determined by the equation

$$x = z^6,$$

which gives $\quad dx = 6z^5 dz.$

The differential expression now reduces to

$$6 \frac{z^8 + z^5}{z^2 + 1} dz.$$

By division this reduces to

$$6(z^6 - z^4 + z^3 + z^2 - z - 1)dz + \frac{6z\,dz}{z^2 + 1} + \frac{6\,dz}{z^2 + 1}.$$

Integrating and replacing z by its equivalent, $x^{\frac{1}{6}}$, we find

$$\int \frac{1 + x^{\frac{1}{2}}}{1 + x^{\frac{1}{3}}} dx = \frac{6}{7} x^{\frac{7}{6}} - \frac{6}{5} x^{\frac{5}{6}} + \frac{6}{4} x^{\frac{4}{6}} + \frac{6}{3} x^{\frac{3}{6}} - \frac{6}{2} x^{\frac{2}{6}} - 6 x^{\frac{1}{6}}$$

$$+ 3 \log (x^{\frac{1}{3}} + 1) + 6 \tan^{(-1)} x^{\frac{1}{6}} + h.$$

If the fractional exponent belongs to a function of x of the first degree, that is, of the form $ax + b$, we apply the same method by substituting the new variable for the proper root of this function.

EXAMPLE. To integrate
$$\frac{(a+bx)^{\frac{1}{2}}dx}{1+(a+bx)}.$$

We put $\quad (a+bx)^{\frac{1}{2}} = z; \quad a+bx = z^2;$

$$dx = \frac{2zdz}{b}.$$

The expression to be integrated now becomes
$$\frac{2z^2 dz}{b(1+z^2)} = \frac{2}{b}\left(dz - \frac{dz}{z^2+1}\right),$$
of which the integral is

$$\frac{2}{b}(z - \tan^{(-1)} z + h) = \frac{2}{b}\left\{(a+bx)^{\frac{1}{2}} - \tan^{(-1)}(a+bx)^{\frac{1}{2}} + h\right\}.$$

EXERCISES.

Integrate:

1. $\dfrac{x^{\frac{1}{2}}dx}{1+x}$.

2. $\dfrac{x^{\frac{1}{2}}dx}{1+x^{\frac{1}{2}}}$.

3. $\dfrac{1-x^{\frac{1}{2}}}{1+x^{\frac{1}{2}}}dx$.

4. $\dfrac{(a-x)^{\frac{1}{2}}dx}{1+a-x}$.

5. $\dfrac{(a-x)^{\frac{1}{2}}dx}{1-(a-x)^{\frac{1}{2}}}$.

6. $\dfrac{1+a-x}{(a-x)^{\frac{1}{2}}}dx$.

7. $\dfrac{(x+c)^{\frac{1}{2}}}{(x+c)^{\frac{1}{2}}}dx$.

8. $\dfrac{(x-c)^{\frac{1}{2}}}{(x-c)^{\frac{1}{2}}}dx$.

9. $\dfrac{(2x-a)^{\frac{1}{2}}dx}{1+(2x-a)^{\frac{1}{2}}}$.

10. $\dfrac{1+(z-c)^{\frac{1}{2}}}{1+(z-c)^{\frac{1}{2}}}dz$.

11. $\dfrac{1-(x+a)^{\frac{1}{2}}}{1+(x+a)^{\frac{1}{2}}}dx$.

12. $\dfrac{1+\dfrac{1}{\sqrt{x}}}{1-\dfrac{1}{\sqrt{x}}}dx$.

13. $\dfrac{x^{\frac{1}{2}}+\dfrac{1}{x^{\frac{1}{2}}}}{x^{\frac{1}{2}}-\dfrac{1}{x^{\frac{1}{2}}}}dx$.

14. $\dfrac{(x-a)^{\frac{1}{2}} - (x-a)^{\frac{1}{2}}}{(x-a)^{\frac{1}{2}} + (x-a)^{\frac{1}{2}}}dx$.

INTEGRATION OF IRRATIONAL DIFFERENTIALS. 235

139. *Cases when the Given Differential contains an Irrational Quantity of the Form*
$$\sqrt{a + bx + cx^2}.$$

It is a fundamental theorem of the Integral Calculus that if we put $R \equiv$ any quadratic function of x, then every expression of the form
$$F(x, \sqrt{R})dx,$$
($F(x, \sqrt{R})$ being a *rational* function of x and $\sqrt{R}$), admits of integration in terms of algebraic, logarithmic, trigonometric or circular functions. But if R contains terms of the third or any higher order in x, then the integral can, in general, be expressed only in terms of certain higher transcendent functions know as elliptic and Abelian functions.

We have three cases of a quadratic function of x.

First Case: c positive. If c is positive, we may render the expression rational by substituting for x the variable z, determined by the equation
$$\sqrt{a + bx + cx^2} = \sqrt{c}(x + z);$$
$$\therefore a + bx + cx^2 = cx^2 + 2cxz + cz^2.$$

This gives $\qquad x = \dfrac{cz^2 - a}{b - 2cz};\qquad$ (a)

$$dx = -2c\frac{a - bz + cz^2}{(b - 2cz)^2}dz; \qquad (b)$$

$$\sqrt{a + bx + cx^2} = -\sqrt{c}\frac{a - bz + cz^2}{b - 2cz}. \qquad (c)$$

By substituting the values given by (a), (b) and (c) for the radical, x, and dx, the expression to be integrated will become rational.

Second Case: a positive and c negative. If the term in x^2 is negative while a is positive, we put
$$\sqrt{a + bx - cx^2} = \sqrt{a} + xz.$$

We thus derive $\qquad x = \dfrac{b - 2\sqrt{a}z}{z^2 + c}; \qquad$ (a)

$$dx = \frac{2(\sqrt{a}z^2 - \sqrt{a c} - bz)}{(z^2 + c)^2} dz;$$

$$\sqrt{a + bx - cx^2} = -\frac{\sqrt{a}z^2 - \sqrt{a c} - bz}{z^2 + c}.$$

The substitution of these expressions will render the equation to be integrated rational.

Third Case: a and c both negative. If the extreme terms of the trinomial are both negative, we find the roots of quadratic equation

$$- a + bx - cx^2 = 0,$$

which roots we call α and β. We then have

$$- a + bx - cx^2 = c(\alpha - x)(x - \beta),$$

and we introduce the new variable z by the condition

$$\sqrt{- a + bx - cx^2} = \sqrt{c(\alpha - x)(x - \beta)} = \sqrt{c}(x - \alpha)z,$$

which gives
$$x = \frac{\alpha z^2 + \beta}{z^2 + 1};$$

$$dx = \frac{2(\alpha - \beta)z\, dz}{(z^2 + 1)^2};$$

$$\sqrt{- a + bx - cx^2} = \frac{\sqrt{c}(\beta - \alpha)z}{z^2 + 1};$$

substitutions which will render the equation rational.

140. We have already integrated one expression of form just considered, namely, $\dfrac{dx}{\sqrt{a + bx + cx^2}}$ without rationalization. There is yet another expression which admits being integrated by a very simple transformation, namely,

$$d\theta = \frac{dr}{r\sqrt{ar^2 + br - 1}}.$$

This is the polar equation of the orbit of a planet around the sun. To integrate it directly, we put

INTEGRATION OF IRRATIONAL DIFFERENTIALS. 237

$$x = \frac{1}{r}; \quad dr = -\frac{dx}{x^2}.$$

We thus reduce the expression to

$$\int \frac{-dx}{\sqrt{a + bx - x^2}}.$$

Proceeding as in § 133, Case I., we find the value of the integral to be

$$\int \frac{dr}{r\sqrt{ar^2 + br - 1}} = \cos^{(-1)} \frac{2x - b}{\sqrt{4a + b^2}} = \cos^{(-1)} \frac{2 - br}{r\sqrt{4a + b^2}}.$$

Thus, $\theta - \pi = \cos^{(-1)} \dfrac{2 - br}{r\sqrt{4a + b^2}}$,

π being an arbitrary constant. Hence

$$\frac{2 - br}{r\sqrt{4a + b^2}} = \cos(\theta - \pi).$$

Solving with respect to r, we have, for the polar equation of the required curve,

$$r = \frac{2}{b + \sqrt{(4a + b^2)} \cos(\theta - \pi)}. \tag{a}$$

This can be readily shown to represent an ellipse. The polar equation of the ellipse is, when the major axis is taken as the base-line and the focus as the pole,

$$r = \frac{a(1 - e^2)}{1 + e \cos \theta} = \frac{2}{\dfrac{2}{a(1 - e^2)} + \dfrac{2e}{a(1 - e^2)} \cos \theta}.$$

Comparing with (a), we have

$$a(1 - e^2) = \frac{2}{b} = \text{parameter of ellipse} \equiv p;$$

$$\frac{2e}{p} = \sqrt{(4a + b^2)},$$

or $\qquad e = \dfrac{\sqrt{(4a + b^2)}}{b} =$ eccentricity of ellipse.

Irrational Binomial Forms.

141. *General Theory.* An irrational binomial differential is one in the form

$$(a + bx^n)^p x^m dx, \qquad (1$$

in which m and n are integers positive or negative, while p i fractional.

To find how and when such a form may be reduced to a rational one, let the fraction p, reduced to its lowest terms, b $\frac{r}{s}$; and let us put

$$y \equiv (a + bx^n)^{\frac{1}{s}}. \qquad (2$$

This will give, when raised to the rth power and multiplied by $x^m dx$,

$$(a + bx^n)^p x^m dx = x^m y^r dx. \qquad (3$$

We readily find, from (2),

$$bx^n = y^s - a; \qquad (a$$

$$dx = \frac{s y^{s-1} dy}{b n x^{n-1}};$$

$$x^m y^r dx = \frac{s}{bn} x^{m-n+1} y^{r+s-1} dy;$$

or, substituting for x its value from (a),

$$x^m y^r dx = \frac{s}{bn} \left(\frac{y^s - a}{b} \right)^{\frac{m-n+1}{n}} y^{r+s-1} dy. \qquad (4$$

This last differential will be rational if $\dfrac{m-n+1}{n}$ is an integer, which will be the case if $\dfrac{m+1}{n}$ is an integer. We shall call this Case I.

To find another case when the integral may be rationalized let us transform the given differential (1) by dividing the binomial by x^n and multiplying the factor outside of it by x^{np} which will leave its value unchanged. It will then be

INTEGRATION OF IRRATIONAL DIFFERENTIALS. 239

$$(b + ax^{-n})^p x^{m + np} dx, \tag{1'}$$

which is another differential of the same form in which n is changed into $-n$ and m into $m + np$. Hence, by Case I., this form can be made rational whenever $\dfrac{m + np + 1}{n}$ is an integer; that is, when $\dfrac{m+1}{n} + p$ is such.

We have, therefore, two cases of integrability, namely:

Case I.: when $\dfrac{m+1}{n}$ = an integer.

Case II.: when $\dfrac{m+1}{n} + p$ = an integer.

REMARK. It will be seen that all differentials of the form $(a + bx^2)^{\frac{r}{2}} x^m dx$ must belong to one of these classes, because $\dfrac{m+1}{2}$ is an integer when m is odd, and $\dfrac{m+1}{2} + \dfrac{r}{2}$ is such when m is even. In this statement we assume r to be odd, because if it is even the original expression is rational.

142. If, in Case I., the integer is $+1$, that is, if $m + 1 = n$, then the expression can be integrated immediately. For (4) then becomes

$$\frac{s}{bn} y^{r + s - 1} dy,$$

the integral of which, after replacing y by its value in (2), becomes

$$\int (a + bx^n)^p x^{n-1} dx = \frac{(a + bx^n)^{p+1}}{bn(p+1)} + c. \tag{5}$$

Again, if the integer in Case II. is -1, we have

$$m + 1 + np = -n,$$

or $\qquad m + np = -n - 1.$

The expression (1) reduced to the form (1') will then be

$$(b + ax^{-n})^p x^{-n-1} dx = -(b + ax^{-n})^p \frac{1}{an} d(b + ax^{-n}),$$

which is immediately integrable, and gives by simple reductions

$$\int (a + bx^n)^p x^{-(np+n+1)} dx = c - \frac{(a+bx^n)^{p+1}}{an(p+1)x^{n(p+1)}}. \quad (6)$$

143. Forms of Reduction of Irrational Binomials. Although the integrable forms can be integrated by the substitution (2), it will, in most cases, be more convenient to apply a system of transformations by which the integrals can be reduced to one of the forms just considered. The objects of these transformations are:

    I. To replace m by $m + n$ or $m - n$;
    II. To replace p by $p + 1$ or $p - 1$.

144. Firstly, to replace m by $m + n$. Let us write, for brevity,

$$X \equiv a + bx^n,$$

which will give $\quad dX = bnx^{n-1} dx,$

and the given differential will be

$$X^p x^m dx,$$

which, again, is equal to

$$\frac{x^{m-n+1}}{bn} X^p dX = \frac{x^{m-n+1}}{bn(p+1)} d(X^{p+1}).$$

Integrating by parts, we have

$$\int X^p x^m dx = \frac{x^{m-n+1} X^{p+1}}{bn(p+1)} - \frac{m-n+1}{bn(p+1)} \int X^{p+1} x^{m-n} dx. \quad (a)$$

Since

$$X^{p+1} = X^p(a + bx^n) = aX^p + bX^p x^n,$$

the last integral in the above equation is the same as

$$a \int X^p x^{m-n} dx + b \int X^p x^m dx,$$

of which the second integral is the same as the original one.
Making this substitution in (a), and then solving the equa-

INTEGRATION OF IRRATIONAL DIFFERENTIALS. 241

tion so as to obtain the value of $\int X^p x^m dx$, we find

$$\int X^p x^m dx = \frac{X^{p+1} x^{m-n+1}}{b(np+m+1)} - \frac{a(m-n+1)}{b(np+m+1)} \int X^p x^{m-n} dx. \quad (A)$$

Thus the given integral is made to depend upon another in which the exponent of x is changed from m to $m-n$.

By reversing the equation we make the given integral depend on one in which the exponent is increased by n. To do this we change m into $m+n$ all through the equation (A), thus getting

$$\int X^p x^{m+n} dx = \frac{X^{p+1} x^{m+1}}{b(np+m+n+1)} - \frac{a(m+1)}{b(np+m+n+1)} \int X^p x^m dx.$$

Solving with respect to the last integral, we find

$$\int X^p x^m dx = \frac{X^{p+1} x^{m+1}}{a(m+1)} - \frac{b(np+m+n+1)}{a(m+1)} \int X^p x^{m+n} dx. \quad (B)$$

The repeated application of (A) and (B) enables us to make the value of the given integral depend upon other integrals of the same form, in which

m is replaced by $\quad m+n; \quad m+2n; \quad$ etc.;

or by $\quad\quad\quad\quad\quad m-n; \quad m-2n; \quad$ etc.

145. Next, to obtain forms in which p is increased or diminished by unity, we express the given differential in the form

$$X^p x^m dx = X^p d\left(\frac{x^{m+1}}{m+1}\right).$$

Integrating by parts and substituting for dX its value $bnx^{n-1} dx$, we have

$$\int X^p x^m dx = \frac{X^p x^{m+1}}{m+1} - \frac{bnp}{m+1} \int X^{p-1} x^{m+n} dx. \quad (b)$$

Now, we have

and therefore, by multiplying by $X^{p-1}dx$,

$$\int X^{p-1}x^{m+n}dx = \frac{1}{b}\int X^p x^m dx - \frac{a}{b}\int X^{p-1}x^m dx.$$

Substituting this value in (b), and solving as before with respect to $\int X^p x^m dx$, we shall find

$$\int X^p x^m dx = \frac{X^p x^{m+1}}{np+m+1} + \frac{anp}{np+m+1}\int X^{p-1}x^m dx, \quad (C)$$

in which p is diminished by unity.

If we write $p+1$ for p in this equation, the last integral will become the given one. Doing this, and then solving with respect to the last integral, we find

$$\int X^p x^m dx = -\frac{X^{p+1}x^{m+1}}{an(p+1)} + \frac{np+n+m+1}{an(p+1)}\int X^{p+1}x^m dx. \quad (D)$$

By the repeated application of the formula (C) or (D) we change

p into $p-1,\ p-2,\ p-3$, etc.,

or $\quad\quad p$ into $p+1,\ p+2,\ p+3$, etc.

146. To see the effect of these transformations, let us put, in the criteria of Cases I. and II., § 141:

$$\text{I.}\quad \frac{m+1}{n} = i, \text{ an integer.}$$

$$\text{II.}\quad \frac{m+1}{n} + p = i', \text{ an integer.}$$

Then when we apply formula (A) or (B), since we replace m by $m-n$ or $m+n$, we have, for the new integers:

$$\text{I.}\quad \frac{m\mp n+1}{n} = i \mp 1.$$

$$\text{II.}\quad \frac{m\mp n+1}{n} + p = i' \mp 1.$$

It is also clear that by (C) and (D) we change II. by unity. Thus, every time we apply formulæ (A), (B), (C) or (D) we change one or both of these integers by unity, so that we may bring them to the values unity treated in § 142.

INTEGRATION OF IRRATIONAL DIFFERENTIALS.

147. *Case of Failure in this Reduction.* If, in an integral of Case II., i' is positive, we cannot change it from zero to -1 by the formula (A) or (C), because, when $\dfrac{m+1}{n} + p = 0$, we have
$$m + 1 + np = 0,$$
and the denominators in (A) and (C) then vanish. In this case we have to apply the substitution of §141, without trying to reduce the integral farther.

EXAMPLES AND EXERCISES.

1. To integrate
$$(a^2 \pm x^2)^{\frac{1}{2}} dx.$$

We see that if we diminish the exponent $\frac{1}{2}$ by unity, we shall reduce the integral to a known elementary form of §132. So we apply (C), putting
$$m = 0; \quad n = 2; \quad p = \tfrac{1}{2}; \quad a = a^2; \quad b = \pm 1.$$
Then (C) becomes
$$\int (a^2 \pm x^2)^{\frac{1}{2}} dx = \frac{x(a^2 \pm x^2)^{\frac{1}{2}}}{2} + \frac{a^2}{2} \int \frac{dx}{(a^2 \pm x^2)^{\frac{1}{2}}}.$$
We therefore have, from §132,
$$\int (a^2 + x^2)^{\frac{1}{2}} dx = \frac{1}{2} \left\{ x(a^2 + x^2)^{\frac{1}{2}} + a^2 \log \frac{C}{a}(x + (a^2 + x^2)^{\frac{1}{2}}) \right\};$$
$$\int (a^2 - x^2)^{\frac{1}{2}} dx = \frac{1}{2} \left\{ x(a^2 - x^2)^{\frac{1}{2}} + a^2 \sin^{(-1)} \frac{x}{a} + h \right\}.$$

Deduce the following equations:

2. $\displaystyle\int (c^2 - x^2)^{\frac{1}{2}} x\, dx \;=\; h - \tfrac{1}{3}(c^2 - x^2)^{\frac{3}{2}}.$

3. $\displaystyle\int (c^2 + x^2)^p x\, dx \;=\; h + \dfrac{(c^2 + x^2)^{p+1}}{2(p+1)}.$

4. $\displaystyle\int (c^2 + x^2)^{\frac{1}{2}} x^{-4} dx \;=\; h - \dfrac{(c^2 + x^2)^{\frac{3}{2}}}{3c^2 x^3}.$

5. $\displaystyle\int \dfrac{dx}{x^2(c^2 + x^2)^{\frac{1}{2}}} \;=\; h - \dfrac{(c^2 + x^2)^{\frac{1}{2}}}{c^2 x}.$

6. $\int \frac{(1-y^4)^{\frac{1}{2}}}{y^7} dy = h - \frac{(1-y^4)^{\frac{3}{2}}}{6y^6}$.

7. $\int (a^2 - x^2)^{\frac{1}{2}} dx = \tfrac{1}{2} x (a^2 - x^2)^{\frac{1}{2}} + \tfrac{1}{2} a^2 \sin^{(-1)} \frac{x}{a}$.

Here apply formula (C); in the following (A).

8. $\int (1 - x^2)^{\frac{1}{2}} x^3 dx = h - \left(\frac{x^2}{5} + \frac{2}{15}\right)(1 - x^2)^{\frac{3}{2}}$.

9. To reduce and integrate $(1 + x^2)^{\frac{1}{2}} x^3 dx$.

Here $m = 3$; $n = 2$; $p = \tfrac{1}{2}$; $m + 1 = 4 = 2n$. We can therefore reduce the form to Case I. by a transformation of m into $m - n$, for which we may use either (a) or (A) of § 144. Using (a), we have

$$\int (1 + x^2)^{\frac{1}{2}} x^3 dx = \frac{(1 + x^2)^{\frac{3}{2}} x^2}{3} - \frac{2}{3}\int (1 + x^2)^{\frac{3}{2}} x dx.$$

The last integral can be immediately found, and gives for the required integral

$$\tfrac{1}{3}(1 + x^2)^{\frac{3}{2}} x^2 - \tfrac{2}{15}(1 + x^2)^{\frac{5}{2}}. \qquad (a)$$

Using (A), we should find

$$\int (1 + x^2)^{\frac{1}{2}} x^3 dx = (1 + x^2)^{\frac{3}{2}} \left(\frac{x^2}{5} - \frac{2}{15}\right), \qquad (b)$$

a form to which (a) can be immediately reduced.

The student will remark that the form (a) is reduced to (A) because in the former the exponent of X is increased by 1, which often makes the integration inconvenient. But when this increase of p does not interfere with the integration, we may use (a) more easily than (A).

10. To reduce and integrate $(1 + x^2)^{\frac{1}{2}} x^5 dx$.

Applying (A), we find

$$\int (1 + x^2)^{\frac{1}{2}} x^5 dx = \frac{(1 + x^2)^{\frac{3}{2}} x^4}{7} - \frac{4}{7}\int (1 + x^2)^{\frac{1}{2}} x^3 dx.$$

A second application repeats the form (b) above, thus giving

$$\int (1 + x^2)^{\frac{1}{2}} x^5 dx = (1 + x^2)^{\frac{3}{2}} \left(\frac{x^4}{7} - \frac{4x^2}{35} + \frac{8}{105}\right).$$

11. Reduce and integrate $(1 + x^2)^{\frac{1}{2}} x^m dx$, where m is any positive odd integer, and show that

$$\int (1+x^2)^{\frac{1}{2}} x^m \, dx$$
$$=(1+x^2)^{\frac{3}{2}}\left(\frac{x^{m-1}}{m+2} - \frac{(m-1)x^{m-3}}{(m+2)m} + \frac{(m-1)(m-3)x^{m-5}}{(m+2)m(m-2)} - \cdots\right).$$

REMARK. Where the student is writing a series of transformations he will find it convenient to put single symbols for the integral expressions which repeat themselves. Thus:

$$\int X^h x^m \, dx \equiv (1); \quad \int X^h x^{m-n} \, dx \equiv (2); \quad \text{etc.}$$

Thus the equations of reduction in the present example may be written

$$(1) = \frac{X^{\frac{3}{2}} x^{m-1}}{m+2} - \frac{m-1}{m+2} \times (2);$$

$$(2) = \frac{X^{\frac{3}{2}} x^{m-3}}{m} - \frac{m-3}{m} \times (3).$$

etc. etc.

12. Deduce the result

$$\int (a^2+x^2)^{\frac{1}{2}} x^6 \, dx = a^2 \left(+x^2 \right)^{\frac{3}{2}} \left(\frac{x^5}{8} - \frac{5a^2 x^3}{48} + \frac{5a^4 x}{64}\right)$$
$$- \frac{5a^6}{128}\left\{ x(a^2+x^2)^{\frac{1}{2}} + a^2 \log C(x + \sqrt{a^2+x^2}) \right\}.$$

CHAPTER V.

INTEGRATION OF TRANSCENDENT FUNCTIONS.

When the given differential contains trigonometric or other transcendent functions of the variable more complex than the simple forms treated in Chapter II., no general method of reduction can be applied. Each case must therefore be studied for itself.

148. To find the integrals

$$\int e^{mx} \cos nx\, dx \quad \text{and} \quad \int e^{mx} \sin nx\, dx. \qquad (1)$$

Since we have

$$e^{mx} dx = \frac{1}{m} d(e^{mx}) = d\!\left(\frac{e^{mx}}{m}\right),$$

the integration by parts of these two expressions gives

$$\int e^{mx} \cos nx\, dx = \frac{e^{mx} \cos nx}{m} + \frac{n}{m}\int e^{mx} \sin nx\, dx;$$

$$\int e^{mx} \sin nx\, dx = \frac{e^{mx} \sin nx}{m} - \frac{n}{m}\int e^{mx} \cos nx\, dx.$$

Solving these equations with respect to the two integrals which they contain, we find

$$\left. \begin{aligned} \int e^{mx} \cos nx\, dx &= \frac{e^{mx}(m \cos nx + n \sin nx)}{m^2 + n^2}; \\ \int e^{mx} \sin nx\, dx &= \frac{e^{mx}(m \sin nx - n \cos nx)}{m^2 + n^2}; \end{aligned} \right\} \qquad (2)$$

which are the required values.

REMARK. These integrals can also be obtained by substituting for the sine and cosine their expressions in terms of imaginary exponentials, namely,

$$2\cos nx = e^{nxi} + e^{-nxi},$$
$$2\sin nx = \frac{1}{i}(e^{nxi} - e^{-nxi}),$$

and then integrating according to the method of § 134. The student should thus deduce the form (2) as an exercise.

149. *Integration of* $\sin^m x \cos^n x dx$.

This form is readily reducible to that of a binomial, and that in two ways. Since we have
$$\cos x dx = d \cdot \sin x,$$
$$\cos x = (1 - \sin^2 x)^{\frac{1}{2}},$$
we see that the integral may be written in the form
$$\int (1 - \sin^2 x)^{\frac{n-1}{2}} \sin^m x d \cdot \sin x;$$
or, putting $y \equiv \sin x$,
$$\int (1 - y^2)^{\frac{n-1}{2}} y^m dy. \tag{3}$$

By putting $z \equiv \cos x$ we should have, in the same way,
$$-\int (1 - z^2)^{\frac{m-1}{2}} z^n dz, \tag{4}$$
which is still of the same form, and is always integrable by the methods already developed in Chapter IV.

If either m or n is a positive odd integer, then by developing the binomial in (3) or (4) by the binomial theorem we shall reduce the expression to a series containing only positive or negative powers of x, which is easily integrable.

We can also, in any case, transform the integral so as to increase or diminish either of the exponents m and n by steps of two units at a time, as follows:
$$\sin^m x \cos^n x dx = \cos^{n-1} x \sin^m x d \cdot \sin x$$
$$= \cos^{n-1} x d \frac{\sin^{m+1} x}{m+1}.$$

Then, integrating by parts, we have

$$\int \sin^m x \cos^n x\,dx$$
$$= \frac{\cos^{n-1} x \sin^{m+1} x}{m+1} + \frac{n-1}{m+1}\int \sin^{m+2} x \cos^{n-2} x\,dx. \quad (5)$$

Because $\sin^{m+2} x = \sin^m x(1 - \cos^2 x)$, the last term is equivalent to

$$\frac{n-1}{m+1}\int \sin^m x \cos^{n-2} x - \frac{n-1}{m+1}\int \sin^m x \cos^n x\,dx.$$

The last of these factors is the original integral. Transposing the term containing it, we find

$$(m+n)\int \sin^m x \cos^n x\,dx = \sin^{m+1} x \cos^{n-1} x$$
$$+ (n-1)\int \sin^m x \cos^{n-2} x\,dx, \quad (6)$$

in which the exponent of $\cos x$ is diminished by 2.

We may in a similar way place the given differential in the form

$$-\sin^{m-1} x\, d\frac{\cos^{n+1} x}{n+1},$$

and then, proceeding as before, we shall find

$$(m+n)\int \sin^m x \cos^n x\,dx = -\sin^{m-1} x \cos^{n+1} x$$
$$+ (m-1)\int \sin^{m-2} x \cos^n x\,dx, \quad (7)$$

thus diminishing the exponent of $\sin x$ by 2.

By reversing these two equations we get forms in which the exponents are increased by 2. Writing $n+2$ for n in the first, and $m+2$ for m in the second, we find

$$(n+1)\int \sin^m x \cos^n x\,dx = -\sin^{m+1} x \cos^{n+1} x$$
$$+ (m+n+2)\int \sin^m x \cos^{n+2} x\,dx; \quad (8)$$

$$(m+1)\int \sin^m x \cos^n x\,dx = \sin^{m+1} x \cos^{n+1} x$$
$$+ (m+n+2)\int \sin^{m+2} x \cos^n x\,dx. \quad (9)$$

150. *Special cases of* $\int \sin^m x \cos^n x\, dx$.

If m is zero and n is positive, we derive, from (6),

$$\left.\begin{array}{l} \int \cos^n x\, dx = \dfrac{\sin x \cos^{n-1} x}{n} + \dfrac{n-1}{n} \int \cos^{n-2} x\, dx; \\[4pt] \int \cos^{n-2} x\, dx = \dfrac{\sin x \cos^{n-3} x}{n-2} + \dfrac{n-3}{n-2} \int \cos^{n-4} x\, dx; \\[2pt] \quad \text{etc.} \qquad\qquad \text{etc.} \qquad\qquad \text{etc.} \end{array}\right\} (10)$$

The integral to be found will thus become that of $\cos x\, dx$ when n is odd, and that of dx, or x itself, when n is even. The given integral is then found by successive substitution.

We find in the same way, from (7),

$$\left.\begin{array}{l} \int \sin^m x\, dx = -\dfrac{\cos x \sin^{m-1} x}{m} + \dfrac{m-1}{m} \int \sin^{m-2} x\, dx; \\[4pt] \int \sin^{m-2} x\, dx = -\dfrac{\cos x \sin^{m-3} x}{m-2} + \dfrac{m-3}{m-2} \int \sin^{m-4} x\, dx; \\[2pt] \quad \text{etc.} \qquad\qquad \text{etc.} \qquad\qquad \text{etc.} \end{array}\right\} (11)$$

From (8) and (9) we derive similar forms applicable to the case of negative exponents.

EXERCISES.

1. $\int \sin^3 x \cos^5 x\, dx.$ *Ans.* $\tfrac{1}{7} \cos^7 x - \tfrac{1}{5} \cos^5 x.$

2. $\int \sin^3 x \cos^2 x\, dx.$ *Ans.* $\tfrac{1}{3} \sin^3 x - \tfrac{1}{5} \sin^5 x.$

3. $\int \dfrac{\cos^3 x\, dx}{\sin^4 x}.$ *Ans.* $\dfrac{3 \sin^2 x - 1}{3 \sin^3 x}.$

4. $\int \sin^2 x \tan^2 x\, dx.$

5. $\int \dfrac{\cos^2 x}{\tan^2 x} dx.$

6. $\int e^{2y} \sin 3y\, dy.$

7. $\int e^{x+a} \cos(x+b)\, dx.$

8. $\int e^{2y} \sin y \cos y\, dy.$

9. $\int e^{-y} \cos^2(y+a)\, dy.$

10. Derive the formulæ of reduction

$$\int \tan^m x\, dx = \dfrac{\tan^{m+1} x}{m+1} - \int \tan^{m+2} x\, dx;$$

and hence

$$\int \tan^n x\,dx = \frac{\tan^{n-1} x}{n-1} - \int \tan^{n-2} x\,dx.$$

These equations may be obtained independently by putting $\tan^n x = \tan^{n-2} x(\sec^2 x - 1)$; or they may be derived from (5).

Hence derive the integrals:

11. $\int \tan^3 x\,dx = \frac{1}{2}\tan^2 x - \log c \sec x.$ (Cf. § 127)

12. $\int \tan^4 x\,dx = \frac{1}{3}\tan^3 x - \tan x + x + h.$

13. For all odd positive integral values of n,

$$\int \tan^n x\,dx = \frac{\tan^{n-1} x}{n-1} - \frac{\tan^{n-3} x}{n-3} + \ldots \pm \log c \sec x.$$

14. When n is positive, integral and an even number,

$$\int \tan^n x\,dx = \frac{\tan^{n-1} x}{n-1} - \frac{\tan^{n-3} x}{n-3} + \ldots \pm \tan x \pm x.$$

15. When the exponent is integral, odd and negative,

$$\int \tan^{-n} x\,dx = -\frac{\cot^{n-1} x}{n-1} + \frac{\cot^{n-3} x}{n-3} - \ldots \pm \log c \sin x.$$

16. When the exponent is integral, even and negative,

$$\int \tan^{-n} x\,dx = -\frac{\cot^{n-1} x}{n-1} + \frac{\cot^{n-3} x}{n-3} - \ldots \pm \cot x \mp x.$$

17. $\int \sin^5 x\,dx = -\frac{\cos x}{5}\left(\sin^4 x + \frac{4}{3}\sin^2 x + \frac{4\cdot 2}{3\cdot 1}\right).$

18. $\int \sin^6 x\,dx$

$$= -\frac{\cos x}{6}\left(\sin^5 x + \frac{5}{4}\sin^3 x + \frac{5\cdot 3}{4\cdot 2}\sin x\right) + \frac{5\cdot 3x}{6\cdot 4\cdot 2}.$$

19. $\int \sin^n x \cos^n x\,dx = \frac{1}{2^n}\int \sin^n 2x\,dx$

151. To integrate $\dfrac{dx}{m^2 \sin^2 x + n^2 \cos^2 x} \equiv du$.

Dividing both terms of the fraction by $\cos^2 x$, noticing that $\dfrac{dx}{\cos^2 x} = d \cdot \tan x$ and writing $t \equiv \tan x$, we find

$$u = \int \dfrac{dt}{m^2 t^2 + n^2}. \tag{12}$$

The integral is known to be (§ 128)

$$\dfrac{1}{mn} \tan^{(-1)} \dfrac{m}{n} t,$$

so that we have

$$u = \int \dfrac{dx}{m^2 \sin^2 x + n^2 \cos^2 x} = \dfrac{1}{mn} \tan^{(-1)} \dfrac{m}{n} \tan x + h, \tag{13}$$

or $\tan x = \dfrac{n}{m} \tan mn(u - h).$

152. Integration of $\dfrac{dy}{a + b \cos y}$.

We reduce this form to the preceding one by the following trigonometric substitution:

$$a = a(\cos^2 \tfrac{1}{2}y + \sin^2 \tfrac{1}{2}y);$$
$$b \cos y = b(\cos^2 \tfrac{1}{2}y - \sin^2 \tfrac{1}{2}y);$$

by which the expression reduces to the form

$$2 \int \dfrac{d(\tfrac{1}{2}y)}{(a - b) \sin^2 \tfrac{1}{2}y + (a + b) \cos^2 \tfrac{1}{2}y}, \tag{14}$$

which is that just integrated, when we put

$$x \equiv \tfrac{1}{2}y;$$
$$m \equiv \sqrt{a - b};$$
$$n \equiv \sqrt{a + b}.$$

We therefore have

$$\int \dfrac{dy}{a + b \cos y} = \dfrac{2}{\sqrt{a^2 - b^2}} \tan^{(-1)} \sqrt{\dfrac{a - b}{a + b}} \tan \tfrac{1}{2}y + h. \tag{15}$$

153. If, in the form of § 151, m^2 and n^2 have opposite signs, or if in § 152 we have $b > a$, imaginary quantities will enter into the integrals, although the latter are real. If, in the first form, the denominator is $m^2 \sin^2 x - n^2 \cos^2 x$, we shall have, instead of (12), the integral

$$\int \frac{dt}{m^2 t^2 - n^2} = \frac{1}{2n} \int \frac{dt}{mt - n} - \frac{1}{2n} \int \frac{dt}{mt + n} \ (\S\ 136)$$

$$= -\frac{1}{2mn} \log \frac{mt + n}{mt - n} + h.$$

Hence, corresponding to (13), we have the result

$$\int \frac{dx}{m^2 \sin^2 x - n^2 \cos^2 x} = -\frac{1}{2mn} \log \frac{m \tan x + n}{m \tan x - n} + h. \quad (16)$$

If, now, in § 152, $b > a$, we write (14) in the form

$$-2\int \frac{d \cdot \tfrac{1}{2}y}{(b - a) \sin^2 \tfrac{1}{2}y - (a + b) \cos^2 \tfrac{1}{2}y},$$

and instead of (15) we have the result

$$\int \frac{dy}{a + b \cos y} = h + \frac{1}{\sqrt{b^2 - a^2}} \log \frac{\sqrt{b-a}\tan\tfrac{1}{2}y + \sqrt{b+a}}{\sqrt{b-a}\tan\tfrac{1}{2}y - \sqrt{b+a}}. \quad (17)$$

154. *Integration of* $\sin mx \cos nx\, dx$.

Every form of this kind is readily integrated by substituting for the products of sines and cosines their expressions in sines and cosines of the sums and differences of the angles. We have, by Trigonometry,

$$\sin mx \cos nx = \tfrac{1}{2} \sin (m + n)x + \tfrac{1}{2} \sin (m - n)x.$$

Hence

$$\int \sin mx \cos nx\, dx = -\frac{\cos (m + n)x}{2(m + n)} - \frac{\cos (m - n)x}{2(m - n)} + h$$

We find in the same way

$$\int \cos mx \cos nx\, dx = \frac{\sin (m + n)x}{2(m + n)} + \frac{\sin (m - n)x}{2(m - n)} + h;$$

$$\int \sin mx \sin nx\, dx = -\frac{\sin (m + n)x}{2(m + n)} + \frac{\sin (m - n)x}{2(m - n)} + h.$$

INTEGRATION OF TRANSCENDENT FUNCTIONS.

155. *Integration by Development in Series.* When the given derived function can be developed in a convergent series, we may find its integral by integrating each term of the series. Of course the integral will then be in the form of a series. The development of many known functions may thus be obtained.

EXAMPLES AND EXERCISES.

1. We may find $\int \sin x \, dx$ as follows: We know that

$$\sin x = x - \frac{x^3}{3!} + \frac{x^5}{5!} - \frac{x^7}{7!} + \ldots ;$$

$$\therefore \int \sin x \, dx = h + \frac{x^2}{2} - \frac{x^4}{4!} + \frac{x^6}{6!} - \text{etc.},$$

which we recognize as the development of $-\cos x$ with an arbitrary constant $h + 1$ added to it.

Of course we may find $\int \cos x \, dx$ in the same way.

2. To integrate $\frac{dx}{1+x}$.

$$\frac{1}{1+x} = (1+x)^{-1} = 1 - x + x^2 - x^3 + \ldots ;$$

$$\therefore \int \frac{dx}{1+x} = h + x - \frac{x^2}{2} + \frac{x^3}{3} - \frac{x^4}{4} + \ldots \qquad (a)$$

Now, we know that $\int \frac{dx}{1+x} = \log(1+x)$. Hence (a) is the development of $\log(1+x)$, when we put $h = \log 1 = 0$.

The series (a) is divergent when $x > 1$. In this case we may form the development by the binomial theorem in descending powers of x, thus:

$$(x+1)^{-1} = x^{-1} - x^{-2} + x^{-3} - x^{-4} + \ldots .$$

Hence we derive, when $x > 1$,

$$\log(x+1) = \log x + \frac{1}{x} - \frac{1}{2x^2} + \frac{1}{3x^3} - \frac{1}{4x^4} + \ldots .$$

The arbitrary constant is zero because, when x is infinite, $\log(x+1) - \log x$ is infinitesimal.

3. To find $\displaystyle\int \frac{dx}{\sqrt{1-x^2}} = \sin^{(-1)} x$ in a series.

$$(1-x^2)^{-\frac{1}{2}} = 1 + \frac{1}{2}x^2 + \frac{1\cdot 3}{2\cdot 4}x^4 + \frac{1\cdot 3\cdot 5}{2\cdot 4\cdot 6}x^6 + \cdots$$

Hence

$$\int \frac{dx}{\sqrt{1-x^2}} = \sin^{(-1)} x = x + \frac{1}{2}\cdot\frac{x^3}{3} + \frac{1\cdot 3}{2\cdot 4}\cdot\frac{x^5}{5} + \frac{1\cdot 3\cdot 5}{2\cdot 4\cdot 6}\cdot\frac{x^7}{7} + \cdots$$

The arbitrary constant is zero by the condition $\sin^{(-1)} 0 = 0$. This series could be used for computing π by putting $x = \frac{1}{2}$, because $\frac{1}{2} = \sin 30° = \sin \frac{\pi}{6}$. But its convergence would be much slower than that of some other series which give the value of π.

4. From the equation $\displaystyle\int \frac{dx}{\sqrt{1+x^2}} = \log(x + \sqrt{1+x^2})$ derive the expansion

$$\log(x + \sqrt{1+x^2}) = x - \frac{1}{2}\cdot\frac{x^3}{3} + \frac{1\cdot 3}{2\cdot 4}\cdot\frac{x^5}{5} - \frac{1\cdot 3\cdot 5}{2\cdot 4\cdot 6}\cdot\frac{x^7}{7} + \cdots$$

5. By expanding $\dfrac{dx}{1+x^2} = d\cdot\tan^{(-1)} x$, derive

$$\tan^{(-1)} x = x - \tfrac{1}{3}x^3 + \tfrac{1}{5}x^5 - \tfrac{1}{7}x^7 + \cdots$$

Derive:

6. $\displaystyle\int \frac{dx}{\sqrt{1+x^4}} = h + x - \frac{1}{2}\cdot\frac{x^5}{5} + \frac{1\cdot 3}{2\cdot 4}\cdot\frac{x^9}{9} - \frac{1\cdot 3\cdot 5}{2\cdot 4\cdot 6}\cdot\frac{x^{13}}{13} + \cdots$

7. $\displaystyle\int e^{-x^2} dx = h + x - \frac{x^3}{3} + \frac{x^5}{5\cdot 2!} - \frac{x^7}{7\cdot 3!} + \cdots$

CHAPTER VI.

OF DEFINITE INTEGRALS.

156. In the Differential Calculus the *increment* of a variable has been defined as the difference between two values of that variable. Let us then suppose u to represent any variable quantity whatever, and let us suppose u to pass through the series of values

$$u_0, u_1, u_2, u_3, \ldots u_n.$$

Then we shall have

$$\Delta u_0 = u_1 - u_0;$$
$$\Delta u_1 = u_2 - u_1;$$
$$\Delta u_2 = u_3 - u_2;$$
$$\cdots$$
$$\Delta u_{n-1} = u_n - u_{n-1}.$$

Taking the sum of all these equations, we have

$$\Delta u_0 + \Delta u_1 + \Delta u_2 + \ldots + \Delta u_{n-1} = u_n - u_0;$$

That is, *the difference between the two extreme values of a variable is equal to the sum of all the successive increments by which it passes from one of these values to the other.*

The same proposition may be shown graphically by supposing the variable to represent the distance from the left-hand end of a line to any point upon the line. The difference between the lengths Δu_0 and Δu_3 is evidently $\Delta u_0 + \Delta u_1 + \ldots + \Delta u_2$.

|---|Δu₀|Δu₁|Δu₂|Δu₃|Δu₄|----
A   u₀ u₁ u₂ u₃ u₄ u₅

Since the proposition is true how small soever the increments, it remains true when they are infinitesimal.

157. *Differential of an Area.* Let P_0PP' be any curve whatever, and let us investigate the differential of the area swept over by the ordinate XP. Let us suppose the foot of the ordinate to start from the position X_0, and move to the position X. During this motion XP sweeps over the area X_0P_0PX, the magnitude of which will depend upon the distance OX, and will therefore be a function of x, which represents this distance.

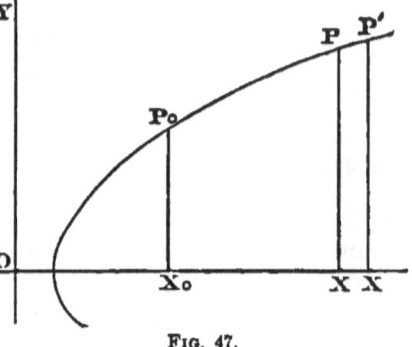

Fig. 47.

Let us put
$u \equiv$ the area swept over;
$y \equiv$ the ordinate XP.

Then, if we assign to x the increment XX', the corresponding increment of the area will be $XPP'X'$. Let us call y' the new ordinate $X'P'$. It is evident that we may always take the increment $XX' \equiv \Delta x$ so small that the area $XPP'X'$ shall be greater than $y\Delta x$ and less than $y'\Delta x$ or *vice versa*. That is, if $y' > y$, as in the figure, we shall have

$$y\Delta x < \Delta u < y'\Delta x.$$

Now, when Δx approaches the limit zero, y' will approach y as its limit, so that the two extremes of this inequality $y\Delta x$ and $y'\Delta x$ will approach equality. Hence, at the limit,

$$du = ydx. \qquad (1)$$

That is, *the area u is such a function of x that its differential is ydx, and its derivative with respect to x is y.*

From this it follows by integration that

$$u = \int ydx + h \qquad (2)$$

is a general expression for the value of the area from any initial ordinate, as X_0P_0 to the terminal ordinate XP.

DEFINITE INTEGRALS.

158. *The Conception of a Definite Integral.* Suppose the area $X_0 P_0 P X \equiv u$ to be divided up into elementary areas, as in the figure. This area will then be made up of the sum of the areas of all the elementary rectangles, plus that of the triangles at the top of the several rectangles.

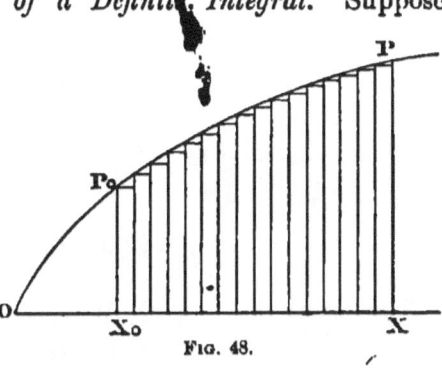

Fig. 48.

That is, using the notation of § 156, we have

$$u = y_0 \Delta x_0 + y_1 \Delta x_1 + y_2 \Delta x_2 + \ldots + y_{n-1} \Delta x_{n-1} + T,$$

T being the sum of the areas of the triangles; or, using the notation of sums,

$$u = \sum_{i=0}^{i=n-1} y_i \Delta x_i + T.$$

Now, let each of the increments Δx_i become infinitesimal. Then each of the small triangles which make up T will become an infinitesimal of the second order, and their sum T will become an infinitesimal of the first order. We may therefore write, for the area u,

$$u = \lim_{x=OX_0}^{x=OX} \sum y \Delta x = \sum_{x=OX_0}^{x=OX} y dx.$$

That is, u is the limit of the sum of all the infinitesimal products ydx, as the foot of the ordinate XP moves from X_0 to X by infinitesimal steps each equal to dx.

Such a sum of an infinite number of infinitesimal products is called a *definite integral*.

The extreme values of the independent variable x, namely, $OX_0 \equiv x_0$ and $OX \equiv x_1$, are called the **limits of integration**.

The infinitesimal increments ydx, whose sum makes up the definite integral, are called its **elements**.

159. Fundamental Theorem. *The definite integral of a continuous function is equal to the difference between the values of the indefinite integral corresponding to the limits of integration.*

To show this let us write $\phi(x)$ for y, and let us put, for the indefinite integral,

$$\int \phi(x)dx = F(x) + c.$$

Now, as already shown, this is a general expression for the area swept over by the ordinate $y = \phi(x)$, when counted from any arbitrary point determined by the constant c. If we count the area from $X_0 P_0$, the area will be zero when $x = x_0$; that is, we must have

$$F(x_0) + c = 0,$$

which gives $\qquad c = - F(x_0).$

If we call x_1 the value of x at X, we shall have

$$u = \text{Area } X_0 P_0 P X = F(x_1) + c = F(x_1) - F(x_0), \quad (3)$$

which was to be proved.

We therefore have a double conception of a definite integral, namely:

(1) As a sum of infinitesimal products;
(2) As the difference between two values of an indefinite integral;

and it will be noticed that the identity of these two conceptions rests on the theorem just enunciated.

Notation. The definite integral is expressed in the same form as the indefinite integral, except that the limits of integration are inserted after the sign $\int$ above and below the line; thus,

$$\int_{x_0}^{x_1} \phi(x)dx$$

means the integral of $\phi(x)dx$ taken between the limits x_0 and x_1, the first being the initial and the second the terminal limit.

Example of the Identity of the Two Conceptions of a Definite Integral. The double conception of a definite integral just reached is of fundamental importance, and may be further illustrated analytically. To take the simplest possible case, consider the definite integral

$$\int_{x_0}^{x_1} a\,dx,$$

a being a constant. By definition this means the sum of all the products

$$a\,dx + a\,dx + a\,dx + \ldots,$$

as x increases from x_0 to x_1. The sum of all the dx's must be equal to $x_1 - x_0$ (§ 156). Hence

$$a(dx + dx + dx + dx + \ldots) = a(x_1 - x_0).$$

But we have for the indefinite integral

$$\int a\,dx = ax;$$

and the definite integral is therefore, by the theorem,

$$ax_1 - ax_0 \quad \text{or} \quad a(x_1 - x_0),$$

as before.

160. *Differentiation of a Definite Integral with respect to its Limits.*—Because the definite integral $\int_{x_0}^{x_1} y\,dx \equiv u$ means the sum of all the products $y\,dx$ as x increases by infinitesimal increments from the lower limit x_0 to the upper limit x_1, or

$$u = y_0 dx + y'dx + y''dx + \ldots + y^{(n)}dx,$$

therefore, assigning an increment dx_1 to the terminal limit x_1 will add the infinitesimal increment $y_1 dx_1$ to u (see Fig. 48). That is, we shall have

$$du = y_1 dx_1, \quad \text{or} \quad \frac{du}{dx_1} = y_1 = \phi(x_1). \qquad (4)$$

In the same way, increasing the initial limit x_0 by dx_0 will take away from the sum the infinitesimal product $y_0 dx_0$, so that we shall have

$$\frac{du}{dx_0} = -y_0 = -\phi(x_0). \tag{5}$$

The equations (4) and (5) give us the derivatives of the definite integral

$$u = \int_{x_0}^{x_1} \phi(x)\,dx$$

with respect to its limits x_1 and x_0.

161. *Examples and Exercises in finding Definite Integrals.*

The fundamental theorem gives the following rule for forming definite integrals:

1. *Form the indefinite integral.*
2. *Substitute for the variable with respect to which we integrate, firstly, the upper limit of integration; secondly, the lower limit.*
3. *Subtract the second result from the first. The difference will be the required definite integral.*

1. $\int_{x_0}^{x_1} x^2 dx = \tfrac{1}{3}x_1^3 - \tfrac{1}{3}x_0^3$.

2. $\int_a^b x\,dx = \tfrac{1}{2}(b^2 - a^2)$.   3. $\int_0^1 x\,dx = \tfrac{1}{2}$.

4. $\int_0^\pi \sin x\,dx = -\cos \pi + \cos 0 = 2$.

5. $\int_0^{\frac{1}{2}\pi} \cos x\,dx = \sin \tfrac{1}{2}\pi$.   6. $\int_b^a az\,dz = \tfrac{1}{2}a(a^2 - b^2)$.

7. $\int_0^{\frac{1}{2}\pi} \sin 2x\,dx$.   8. $\int_{45°}^{90°} \cos 2x\,dx$.

9. $\int_0^{2\pi} \sin^2 x\,dx$.   10. $\int_0^{2\pi} \cos^2 x\,dx$.

11. $\int_0^\pi x \sin x\,dx$.   12. $\int_0^\pi z \cos z\,dz$.

13. $\int_0^\pi z^2 \sin z\,dz$.   14. $\int_0^\pi z^2 \cos z\,dz$.

DEFINITE INTEGRALS. 261

15. $\int_0^{2\pi} z^2 \cos 2z\,dz.$

16. $\int_0^{2\pi} z^2 \sin 2z\,dz.$

17. $\int_1^n \frac{dx}{x}.$

18. $\int_1^a nz^n\,dz.$

19. $\int_{-b}^{b} \frac{dx}{x^2 - a^2}.$

20. $\int_x^{x^2} \frac{dz}{z^2 - 1}.$

21. $\int_{-\pi}^{+\pi} \cos x\,dx.$

22. $\int_{-\pi}^{+\pi} \sin x\,dx.$

23. $\int_{+2}^{+3} \frac{dz}{z^2 - 1}.$

24. $\int_{-1}^{+1} (z - 1)^2\,dz.$

25. $\int_{a-b}^{a+b} (x - a)\,dx.$

26. $\int_{a-x}^{a+x} y\,dy.$

27. $\int_{1+x}^{1-x} (x - 1)^2\,dx.$

28. $\int_{a-c}^{a+c} (x - a)(x - c)\,dx.$

29. $\int_a^b y\,dy + \int_b^a x\,dx.$

30. $\int_{-1}^{+1} \frac{dz}{1 + z^2} = \frac{\pi}{2}.$

31. $\int_0^\pi \sin ax\,dx.$

32. $\int_0^{\frac{\pi}{2}} \cos(a + x)\,dx.$

33. Deduce $\int_{-y}^{+y} \cos(x + y)\,dx = \sin 2y.$

34. Show that $\int_a^b f(x)\,dx = -\int_b^a f(x)\,dx.$

35. Deduce $\int_0^\infty e^{-y}\,dy = 1.$

36. Deduce $\int_0^\infty e^{-ay}\,dy = \frac{1}{a}.$

37. Deduce $\int_{-\infty}^0 e^y\,dy = 1.$

38. Deduce $\int_0^\infty e^{-y^2} y\,dy = \tfrac{1}{2}.$

39. Deduce $\int_{-\infty}^{+\infty} \frac{dz}{1 + z^2} = \pi.$

40. Deduce $\int_0^1 \frac{dz}{\sqrt{1 - z^2}} = \frac{\pi}{2}.$

41. Deduce $\int_{-a}^{+a} \frac{dz}{\sqrt{a^2 - z^2}} = \pi.$

162. *Failure of the Method when the Function becomes Infinite.* It is to be noted that the equivalence of the two conceptions of a definite integral does not necessarily hold true unless the function y or $\phi(x)$ is continuous and finite between the limits of integration. As an example of the failure of this condition, consider the function

$$y = \frac{1}{(x-a)^2},$$

the curve representing which is shown in the margin.

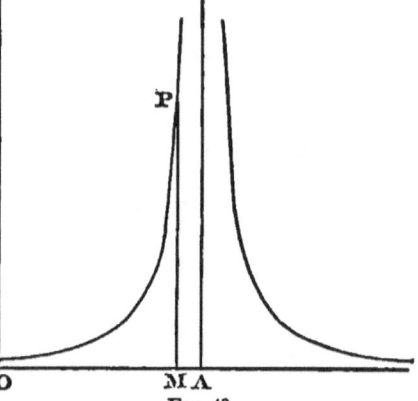

Fig. 49.

The indefinite integral is

$$u = \int y\,dx = c - \frac{1}{x-a}.$$

To find the value of this integral between two such limits as 0 and k, k being any quantity OM less than a, we put $x = 0$ and $x = k$, and take the difference as usual. Thus

$$u = \frac{1}{a-k} - \frac{1}{a} = \frac{k}{a(a-k)}. \tag{5}$$

Now, if we suppose k to approach a as its limit, so that $a - k$ shall become infinitesimal, then the area u will increase without limit, as we readily see from the figure as well as by the formula.

But suppose $k > a$; for example, $k = 2a$. Then the theorem would give

$$\int_0^{2a} y\,dx = -\frac{1}{a} - \frac{1}{a} = -\frac{2}{a},$$

a negative finite quantity; whereas, in reality, the area is an infinite quantity.

The theorem fails because, when $x = a$, y becomes infinite, so that $y\,dx$ is not then necessarily an infinitesimal, as is presupposed in the demonstration.

163. *Change of Variable in Definite Integrals.* When, in order to integrate an expression, we introduce a new variable, we must assign to the limits of integration the values of the new variable which correspond to the limiting values of the old one. Some examples will make this clear.

Ex. 1. Let the definite integral be

$$\int_0^a \frac{dx}{a+x}.$$

Proceeding in the usual way, we find the indefinite integral to be $\log(a+x)$, whence we conclude

$$\int_0^a \frac{dx}{a+x} = \log 2a - \log a = \log 2.$$

But suppose that we transformed the integral by putting

$$y \equiv a + x; \quad dy = dx.$$

Since, at the lower limit, $x = 0$, we must then have $y = a$ for this limit, and when, at the upper limit, $x = a$, we have $y = 2a$. Hence the transformed integral is

$$\int_a^{2a} \frac{dy}{y},$$

which we find to have the same value, $\log 2$.

Ex. 2. $u = \int_0^{\frac{\pi}{2}} \sin x (1 - \cos x) dx.$

We may write the indefinite integral in the form

$$\int \sin x \, dx + \int \cos x \, d(\cos x).$$

In the first term x is still the independent variable. But, as the second is written, $\cos x$ is the independent variable. Now, for

$$x = 0, \quad \cos x = 1;$$

and for

$$x = \frac{\pi}{2}, \quad \cos x = 0.$$

Hence, writing y for $\cos x$, the value of u is

$$u = \int_0^{\frac{\pi}{2}} \sin x \, dx + \int_1^0 y \, dy = 1 - \tfrac{1}{2} = \tfrac{1}{2}.$$

REMARK. *The variable with respect to which the integration is performed always disappears from the definite integral,* which is a function of the limits of integration, and of any quantities which may enter into the differential expression. Hence we may change the symbol of the variable at pleasure without changing the integral. Thus whatever be the form of the function ϕ, or the original meaning of the symbols x and y, we shall always have

$$\int_a^b \phi(x)dx = \int_a^b \phi(y)dy = \int_0^{b-a} \phi(y+a)dy, \text{ etc.}$$

164. *Subdivision of a Definite Integral.* The following definitions come into use here:

1. An *even function* of x is a function whose value remains unchanged when x changes its sign.

2. An *odd function* of x is one which retains the same absolute value with the opposite sign when x changes its sign.

As examples: cos x is an even, sin x an odd, function.

Any function of x^2 is even; the product of any even function into x is odd.

It is evident, from the nature and formation of a definite integral, that if we have a sum of such integrals,

$$\int_a^b \phi(x)dx + \int_b^c \phi(x)dx + \int_c^d \phi(x)dx + \ldots + \int_g^h \phi(x)dx,$$

in which the upper limit of each integral is the lower limit of that next following, this sum is equal to

$$\int_a^h \phi(x)dx.$$

This theorem may often be applied to simplify the expression of the integral in cases where the values of $\phi(x)$ repeat themselves.

THEOREM I. *If $\phi(x)$ is an even function of x, then, whatever be a,*

$$\int_{-a}^{+a} \phi(x)dx = 2\int_0^a \phi(x)dx.$$

Because $\phi(-x) = \phi(x)$, it follows that for every negative value of x between $-a$ and 0 the element of $\phi(x)dx$ will be the same as for the corresponding positive value of x. Hence the infinitesimal sums which make up the value of $\int_{-a}^{0} \phi(x)dx$ will be equal to those which make up $\int_{0}^{a} \phi(x)dx$. Therefore

$$\int_{-a}^{a} \phi(x)dx = \int_{-a}^{0} \phi(x)dx + \int_{0}^{a} \phi(x)dx = 2\int_{0}^{a} \phi(x)dx.$$

THEOREM II. *If $\phi(x)$ is an odd function of x, then, whatever be a,*

$$\int_{-a}^{+a} \phi(x)dx = 0.$$

For in this case each element $\phi(-x)dx$ will be the negative of the element $\phi(x)dx$, and thus the positive and negative elements will cancel each other.

EXERCISES.

1. Show that $\int_{0}^{\infty} e^{-x}x^n dx = \int_{0}^{1} \left(\log \frac{1}{z}\right)^n dz.$

Substitute $x = \log \frac{1}{z}$.

2. Show that whatever be the function ϕ, we have

$$\int_{0}^{\frac{1}{2}\pi} \phi(\sin z)dz = \int_{0}^{\frac{1}{2}\pi} \phi(\cos x dx).$$

As an example of this theorem,

$$\int_{0}^{\frac{1}{2}\pi} \frac{a + b \cos^n x}{a - b \cos^n x} dx = \int_{0}^{\frac{1}{2}\pi} \frac{a + b \sin^n x}{a - b \sin^n x} dx.$$

The truth of this theorem may be seen by showing that to each element of the one integral corresponds an equal element of the other. Draw two quadrants; draw a sine in one and an equal cosine in the other. Any function ϕ of the sine is equal to the corresponding function of the cosine. We may fill one quadrant up with sines and the other with cosines equal to those sines, and then the two integrals will be made up of equal elements.

To express this proof analytically, we replace x by a new variable $y = \frac{1}{2}\pi - x$, which gives $\sin x = \cos y$; $dx = -dy$; and then we invert the limits of the transformed integral, and change y into x in accordance with the remark of the last article.

3. Show that $\displaystyle\int_0^\pi f(\sin x)dx = 2\int_0^{\frac{\pi}{2}} f(\sin x)dx$.

4. Show that $\displaystyle\int_0^\pi \phi(\sin x)\cos x\, dx = 0$.

5. Show that if ϕ be an odd function, then
$$\int_0^\pi \phi(\cos x)dx = 0.$$

6. Show that the product of two like functions, odd or even, is an even function, and that the product of an even and an odd function is an odd function.

7. Show that when ϕ is an odd function, $\phi(0) = 0$.

165. *Definite Integrals through Integration by Parts.*—In the formula for integration by parts, namely,
$$\int u\,dv = uv - \int v\,du,$$
let us apply the rule for finding the definite integral. To express the result, let us put

$(uv)_1$ and $(uv)_0$, the values of uv for the upper and lower limits of integration, respectively;

$\displaystyle\int^{x_1} u\,dv$ and $\displaystyle\int^{x_1} v\,du$, the values of the two indefinite integrals for the upper limit, x_1;

$\displaystyle\int_{x_0} u\,dv$ and $\displaystyle\int_{x_0} v\,du$, the values of the integrals for the lower limit, x_0.

We then have, by the rule of § 161,
$$\int_{x_0}^{x_1} u\,dv = \int^{x_1} u\,dv - \int_{x_0} u\,dv$$
$$= (uv)_1 - \int^{x_1} v\,du - (uv)_0 + \int_{x_0} v\,du$$
$$= (uv)_1 - (uv)_0 - \int_{x_0}^{x_1} v\,du.$$

DEFINITE INTEGRALS.

In order to assimilate the form of this expression to that of a definite integral, it is common to write

$$(uv)'_0 \equiv (uv)_1 - (uv)_0.$$

EXAMPLES AND EXERCISES.

1. We have found the indefinite integral

$$\int \log x\, dx = x \log x - \int dx.$$

If we take this integral between the limits $x = 0$ and $x = 1$, the term $x \log x$ will vanish at both limits, so that

$$(x \log x)_1 - (x \log x)_0 = 0.$$

Hence $\int_0^1 \log x\, dx = -\int_0^1 dx = -1 + 0 = -1.$

2. To find the definite integral,

$$\int_0^\pi \sin^m x\, dx.$$

In the equation (11), § 150, the first term of the second member vanishes at both the limits $x = 0$ and $x = \pi$. Hence

$$\int_0^\pi \sin^m x\, dx = \frac{m-1}{m} \int_0^\pi \sin^{m-2} x\, dx.$$

Writing $m - 2$ for m, and repeating the process, we have

$$\int_0^\pi \sin^{m-2} x\, dx = \frac{m-3}{m-2} \int_0^\pi \sin^{m-4} x\, dx;$$

$$\int_0^\pi \sin^{m-4} x\, dx = \frac{m-5}{m-4} \int_0^\pi \sin^{m-6} x\, dx;$$

etc.   etc.

If m is even, we shall at length reach the form

$$\int_0^\pi dx = \pi - 0 = \pi.$$

Then, by successive substitution, we shall have

$$\int_0^\pi \sin^m x\,dx = \frac{(m-1)(m-3)(m-5)\ldots 1}{m(m-2)(m-4)\ldots 2}\cdot \pi.$$

If m is odd, the last integral will be $\int_0^\pi \sin x\,dx = +2$, and we shall have

$$\int_0^\pi \sin^m x\,dx = 2\frac{(m-1)(m-3)\ldots 2}{m(m-2)(m-4)\ldots 3}.$$

3. From the equation (6) of § 149 we have, by forming the definite integral and dividing by $m+n$,

$$\int_0^\pi \sin^m x \cos^n x\,dx = \left(\frac{\sin^{m+1} x \cos^{n-1} x}{m+n}\right)_0^\pi$$
$$+ \frac{n-1}{m+n}\int_0^\pi \sin^m x \cos^{n-2} x\,dx.$$

Since $\sin \pi = \sin 0 = 0$, the first term of the second member vanishes between the limits, and we have

$$\int_0^\pi \sin^m x \cos^n x\,dx = \frac{n-1}{m+n}\int_0^\pi \sin^m x \cos^{n-2} x\,dx.$$

Writing $n-2$, and then $n-4$, etc., in place of n, this formula becomes

$$\int_0^\pi \sin^m x \cos^{n-2} x\,dx = \frac{n-3}{m+n-2}\int_0^\pi \sin^m x \cos^{n-4} x\,dx;$$

$$\int_0^\pi \sin^m x \cos^{n-4} x\,dx = \frac{n-5}{m+n-4}\int_0^\pi \sin^m x \cos^{n-6} x\,dx;$$

etc.   etc.

If n is odd, the successive applications of this substitution will at length lead us to the form

$$\int_0^\pi \sin^m x \cos x\,dx = \frac{1}{m+1}(\sin^{m+1} \pi - \sin^{m+1} 0) = 0;$$

and thus, by successive substitution, we shall find all the integrals to be zero,

If n is even, we shall be led to the form

$$\int_0^\pi \sin^m x\, dx,$$

which we have just integrated. Then, by successive substitution, we find

$$\int_0^\pi \sin^m x \cos^n x$$
$$= \frac{(n-1)(n-3)\ldots 1}{(m+n)(m+n-2)\ldots(m+2)} \int_0^\pi \sin^m x\, dx.$$

4. To find $\int_0^\infty \frac{dx}{(a^2+x^2)^n}$.

We transform the differential thus:

$$\int \frac{dx}{(x^2+a^2)^n} = \frac{1}{a^2} \cdot \int \frac{(x^2+a^2-x^2)\, dx}{(x^2+a^2)^n}$$
$$= \frac{1}{a^2} \cdot \int \frac{dx}{(x^2+a^2)^{n-1}} - \frac{1}{a^2} \int \frac{x^2 dx}{(x^2+a^2)^n}. \quad (a)$$

Integrating the last term by parts, we have

$$\int \frac{x^2 dx}{(x^2+a^2)^n} = \frac{1}{2} \int x \cdot \frac{2x\, dx}{(a^2+x^2)^n} = \frac{1}{2} \int x \frac{d\cdot(x^2+a^2)}{(x^2+a^2)^n}$$
$$= \frac{1}{2} \int x d \cdot \frac{\frac{1}{(x^2+a^2)^{n-1}}}{1-n} = -\frac{1}{2(n-1)} \cdot \frac{x}{(x^2+a^2)^{n-1}}$$
$$+ \frac{1}{2(n-1)} \int \frac{dx}{(x^2+a^2)^{n-1}}.$$

Substituting this value of the last term in (a), we have

$$\int \frac{dx}{(x^2+a^2)^n} = \frac{1}{2a^2(n-1)} \cdot \frac{x}{(x^2+a^2)^{n-1}}$$
$$+ \frac{1}{a^2}\left(1 - \frac{1}{2(n-1)}\right) \int \frac{dx}{(x^2+a^2)^{n-1}}.$$

Passing now to the limits, we see that the first term of the second member vanishes both for $x=0$ and for $x=\infty$. We also have

$$1 - \frac{1}{2(n-1)} = \frac{2n-3}{2(n-1)}.$$

Hence we have the formula of reduction

$$\int_0^\infty \frac{dx}{(x^2+a^2)^n} = \frac{2n-3}{2(n-1)a^2}\int_0^\infty \frac{dx}{(a^2+x^2)^{n-1}}. \quad (b)$$

We can thus diminish the exponent by successive steps until it reaches 2. The formula (b) will then give

$$\int_0^\infty \frac{dx}{(x^2+a^2)^2} = \frac{1}{2a^2}\int_0^\infty \frac{dx}{a^2+x^2} = \frac{\pi}{4a^3}.$$

Then, by successive substitution in the form (b), we shall have

$$\int_0^\infty \frac{dx}{(x^2+a^2)^n} = \frac{(2n-3)(2n-5)\ldots 1}{(2n-2)(2n-4)\ldots 2}\cdot\frac{\pi}{2a^{2n-1}}. \quad (c)$$

If in (c) we suppose $a=1$, and write the second member in reverse order, we have

$$\int_0^\infty \frac{dx}{(1+x^2)^n} = \frac{1\cdot 3\cdot 5\ldots (2n-3)}{2\cdot 4\cdot 6\ldots (2n-2)}\cdot\frac{\pi}{2}.$$

5. To find $\int_0^1 \frac{x^m dx}{\sqrt{1-x^2}} \equiv y_m.$

Let us apply to the indefinite integral the formula (A), § 144. We have in this case

$$a=1; \quad b=-1; \quad n=2; \quad p=-\tfrac{1}{2}.$$

The formula then becomes

$$\int \frac{x^m dx}{\sqrt{1-x^2}} = -\frac{x^{m-1}\sqrt{1-x^2}}{m} + \frac{m-1}{m}\int \frac{x^{m-2}dx}{\sqrt{1-x^2}}. \quad (a)$$

In the same way

$$\int \frac{x^{m-2}dx}{\sqrt{1-x^2}} = -\frac{x^{m-3}\sqrt{1-x^2}}{m-2} + \frac{m-3}{m-2}\int \frac{x^{m-4}dx}{\sqrt{1-x^2}}.$$

Continuing the process, we shall reduce the exponent of x to 1 if m is odd, or to 0 if m is even. Then we shall have

$$\int \frac{x\,dx}{\sqrt{1-x^2}} = -(1-x^2)^{\frac{1}{2}} \quad \text{or} \quad \int \frac{dx}{\sqrt{1-x^2}} = \sin^{(-1)}x. \quad (b)$$

Taking the several integrals between the limits 0 and 1, we

note that in (a) the first term of the second member vanishes at both limits, while (b) gives

$$\int_0^1 \frac{xdx}{\sqrt{1-x^2}} = 1; \quad \int_0^1 \frac{dx}{\sqrt{1-x^2}} = \frac{1}{2}\pi.$$

We thus have, by successive substitution,

$$\left. \begin{array}{l} y_{2n+1} \equiv \int_0^1 \dfrac{x^{2n+1}dx}{\sqrt{1-x^2}} = \dfrac{2n(2n-2)(2n-4)\ldots 2}{(2n+1)(2n-1)(2n-3)\ldots 3}; \\[2mm] y_{2n} \equiv \int_0^1 \dfrac{x^{2n}dx}{\sqrt{1-x^2}} = \dfrac{(2n-1)(2n-3)(2n-5)\ldots 1}{2n(2n-2)(2n-4)\ldots 2}\cdot\dfrac{\pi}{2}. \end{array} \right\} \quad (c)$$

Let us now consider the limit toward which the ratio of two values of y_m approaches as m increases to infinity. We find, from (a),

$$\frac{y_m}{y_{m-2}} = \frac{m-1}{m},$$

a ratio of which unity is the limit.

Next we find, by taking the quotient of the equations (c),

$$\frac{\pi}{2} = \frac{\{2\cdot 4\cdot 6 \ldots (2n-2)\cdot 2n\}^2}{\{3\cdot 5\cdot 7 \ldots (2n-1)\}^2(2n+1)} \cdot \frac{y_{2n}}{y_{2n+1}}.$$

Since, when n becomes infinite, the ratio $y_{2n} : y_{2n+1}$ approaches unity as its limit, we conclude that $\frac{1}{2}\pi$ may be expressed in the form of an infinite product, thus:

$$\frac{\pi}{2} = \frac{4}{3}\cdot\frac{4^2}{3\cdot 5}\cdot\frac{6^2}{5\cdot 7}\cdot\frac{8^2}{7\cdot 9}\cdot\frac{10^2}{9\cdot 11} \ldots \textit{ad infinitum}.$$

This is a celebrated expression for π, known as Wallis's formula. It cannot practically be used for computing π, owing to the great number of factors which would have to be included.

CHAPTER VII.

SUCCESSIVE INTEGRATION.

166. *Differentiation under the Sign of Integration.* Let us have an indefinite integral of the form

$$u = \int \phi(\alpha, x)dx \equiv F(\alpha, x), \qquad (1)$$

α being any quantity whatever independent of x. It is evident that u will in general be a function of α. We have now to find the differential of u with respect to α.

The differentiation of (1) gives

$$\frac{du}{dx} = \phi(\alpha, x);$$

$$\frac{d^2u}{d\alpha dx} = \frac{d\phi(\alpha, x)}{d\alpha}.$$

Because $\dfrac{d^2u}{d\alpha dx} = D_\alpha \dfrac{du}{dx} = D_x \dfrac{du}{d\alpha}$, we have, when we consider $\dfrac{du}{d\alpha}$ as a function of x (cf. § 51),

$$d\left(\frac{du}{d\alpha}\right) = \frac{d^2u}{dx d\alpha}dx = \frac{d\phi(\alpha, x)}{d\alpha}dx.$$

Then, by integrating with respect to x,

$$\frac{du}{d\alpha} = \int \frac{d\phi(\alpha, x)}{d\alpha}dx, \qquad (2)$$

in which the second member is the same as (1), except that $\phi(\alpha, x)$ is replaced by its derivative with respect to α. Hence we have the theorem:

The derivative of an integral with respect to any quantity which enters into it is expressed by differentiating with respect to that quantity under the sign of integration.

167. This theorem being proved for an indefinite integral, we have to inquire whether it can be applied to a definite integral. If we take the integral (1) between the limits x_0 and x_1, and put u_0 and u_1 for the corresponding values of u, we have, for the definite integral,

$$\int_{x_0}^{x_1} \phi(\alpha, x)dx = F(\alpha, x_1) - F(\alpha, x_0) = u_1 - u_0 \equiv u_0'.$$

Then, by differentiation,

$$\frac{du_0'}{d\alpha} = \frac{dF(\alpha, x_1)}{d\alpha} - \frac{dF(\alpha, x_0)}{d\alpha} \qquad (3)$$

Comparing (1) and (2), we have

$$\int \frac{d\phi(\alpha, x)}{d\alpha} dx = \frac{dF(\alpha, x)}{d\alpha};$$

whence, if x_1 and x_0 are *not* functions of α,

$$\int_{x_0}^{x_1} \frac{d\phi(\alpha, x)}{d\alpha} dx = \frac{dF(\alpha, x_1)}{d\alpha} - \frac{dF(\alpha, x_0)}{d\alpha}. \qquad (4)$$

Hence from (3) we have the general theorem

$$D_\alpha \int_{x_0}^{x_1} \phi(\alpha, x)dx = \int_{x_0}^{x_1} D_\alpha \phi(\alpha, x)dx.$$

That is, *the symbols of differentiation and integration with respect to two independent quantities may be interchanged in a definite integral, provided that the limits of integration are not functions of the quantity with respect to which we differentiate.*

If the limits x_1 and x_0 are functions of α, we have, for the total derivative of u_0' with respect to α (§ 41),

$$\frac{du_0'}{d\alpha} = \left(\frac{du_0'}{d\alpha}\right) + \frac{du_0'}{dx_1}\frac{dx_1}{d\alpha} + \frac{du_0'}{dx_0}\frac{dx_0}{d\alpha}.$$

By § 160 we have

$$\frac{du_0'}{dx_1} = \phi(\alpha, x_1);$$
$$\frac{du_0'}{dx_0} = -\phi(\alpha, x_0).$$

Thus from (3) and (4) we have

$$\frac{du_0{}^1}{d\alpha} = \int_{x_0}^{x_1} \frac{d\phi(\alpha, x)}{d\alpha} dx + \phi(\alpha, x_1)\frac{dx_1}{d\alpha} - \phi(\alpha, x_0)\frac{dx_0}{d\alpha}. \quad (5)$$

This formula is subject to the same restriction as the theorem for the value of a definite integral; that is, $\phi(\alpha, x)$ and its derivative with respect to α must be finite and continuous for all values of x between the limits of integration.

If this condition is not fulfilled, (5) may fail.

EXERCISES.

Differentiate:

1. $\int \dfrac{dx}{x + \alpha}$ with respect to α. Ans. $-\int \dfrac{dx}{(x + \alpha)^2}$.

2. $\int (x + \alpha)^n dx$ with respect to α. Ans. $n \int (x+\alpha)^{n-1} dx$.

3. $\int (x^2 + xy)^2 dx$ with respect to y. Ans. $2 \int (x^3 + x^2 y) dx$.

4. $\int_0^\alpha x^2 dx$ with respect to α. Ans. α^2.

5. $\int_0^{2\alpha} x^2 dx$ with respect to α. Ans. $8\alpha^2$.

6. $\int_\alpha^{\alpha^2} x^n dx$ with respect to α. Ans. $= \alpha^n(2\alpha^{n+1} - 1)$.

And show that we have the same results in the first three cases whether we integrate the differential with respect to α or y, or differentiate the integral.

168. The preceding method enables us to find many integrals, indefinite and definite, by differentiating known integrals with respect to constants which enter into them. Thus, by differentiating with respect to a the integral

$$\int e^{ax} dx = \frac{1}{a} e^{ax} + c,$$

we find, after adding the constants of integration,

$$\int x e^{ax} dx = \left(\frac{x}{a} - \frac{1}{a^2}\right) e^{ax} + c;$$
$$\int x^2 e^{ax} dx = \left(\frac{x^2}{a} - \frac{2x}{a^2} + \frac{2}{a^3}\right) e^{ax} + c;$$
etc.　　　　etc.

which leads to the same results as integration by parts, and is shorter.

169. The following is an instructive application of this and other principles. We shall hereafter show that

$$\int_{-\infty}^{+\infty} e^{-x^2} dx = \sqrt{\pi}.$$

From this it is required to find the value of $\int_{-\infty}^{+\infty} e^{-a^2 y^2} dy$. If we put

$$x \equiv ay,$$

whence $$dy = \frac{dx}{a},$$

the corresponding indefinite integral will be

$$\int e^{-a^2 y^2} dy = \frac{1}{a} \int e^{-x^2} dx.$$

Now, when $y = \pm \infty$, we have also $x = \pm \infty$. Hence

$$\int_{-\infty}^{+\infty} e^{-a^2 y^2} dy = \frac{1}{a} \int_{-\infty}^{+\infty} e^{-x^2} dx = \frac{\sqrt{\pi}}{a}.$$

By differentiating with respect to a, and simple reductions, we find

$$\int_{-\infty}^{+\infty} y^2 e^{-a^2 y^2} dy = \frac{\sqrt{\pi}}{2a^3};$$

and from this,

EXERCISES.

1. By differentiating the integrals

$$\int \cos ax\, dx = \frac{1}{a} \sin ax,$$

$$\int \sin ax\, dx = -\frac{1}{a} \cos ax,$$

twice with respect to a, prove the formulæ

$$\int x^2 \cos ax\, dx = \left(\frac{x^2}{a} - \frac{2}{a^3}\right) \sin ax + \frac{2x}{a^2} \cos ax;$$

$$\int x^2 \sin ax\, dx = \left(\frac{2}{a^3} - \frac{x^2}{a}\right) \cos ax + \frac{2x}{a^2} \sin ax.$$

Thence show that we have

$$\int y^2 \cos y\, dy = (y^2 - 2) \sin y + 2y \cos y;$$

$$\int y^2 \sin y\, dy = (2 - y^2) \cos y + 2y \sin y.$$

2. Prove the formulæ:

(a) $\int_{-\infty}^{0} e^{ax}\, dx = \frac{1}{a};$ (b) $\int_{-\infty}^{0} xe^{ax}\, dx = -\frac{1}{a^2};$

(c) $\int_{-\infty}^{0} x^2 e^{ax}\, dx = \frac{2}{a^3};$ (d) $\int_{-\infty}^{0} x^n e^{ax}\, dx = (-1)^n \frac{n!}{a^n}.$

3. Show that the preceding formulæ are true only when a is positive, and find the following corresponding forms when a has the negative sign:

$$\int_{0}^{\infty} e^{-ax}\, dx = \frac{1}{a}; \quad \int_{0}^{\infty} xe^{-ax}\, dx = \frac{1}{a^2};$$

$$\int_{0}^{\infty} x^2 e^{-ax}\, dx = \frac{2}{a^3}; \quad \int_{0}^{\infty} x^3 e^{-ax}\, dx = \frac{2\cdot 3}{a^4}; \text{ etc.}$$

4. By differentiating the form of § 132, namely,

$$\int \frac{dx}{(a^2 - x^2)^{\frac{1}{2}}} = \sin^{(-1)} \frac{x}{a},$$

with respect to a, show that

$$\int \frac{dx}{(a^2 - x^2)^{\frac{3}{2}}} = \frac{x}{a^2(a^2 - x^2)^{\frac{1}{2}}}.$$

170. *Double Integrals.* The preceding results may be summed up and proved thus: Let us have an integral of the form

$$u = \int \phi(x, y)dx, \qquad (1)$$

and let us consider the integral

$$\int u\,dy \quad \text{or} \quad \int \left[\int \phi(x, y)dx\right]dy,$$

which, for brevity, is written without brackets, thus:

$$\int\int \phi(x, y)dx dy.$$

This expression is called a *double integral*.

THEOREM. *The value of an indefinite double integral remains unchanged when we change the order of the integrations, provided that we assign suitable values to the arbitrary constants of integration.*

Let us put

$$v = \int \phi(x, y)dy,$$

u retaining the value (1). The theorem asserts that

$$\int u\,dy = \int v\,dx.$$

Call these two quantities U and V, respectively. We then have, by differentiation,

$$\frac{dU}{dy} = u; \quad \frac{d^2U}{dx\,dy} = \frac{du}{dx} = \phi(x, y);$$

$$\frac{dV}{dx} = v; \quad \frac{d^2V}{dy\,dx} = \frac{dv}{dy} = \phi(x, y).$$

Therefore, because of the interchangeability of differentiations,

$$\frac{d \cdot \frac{dU}{dx}}{dy} = \frac{d \cdot \frac{dV}{dx}}{dy}.$$

Then, by integration with respect to y,

$$\frac{dU}{dx} = \frac{dV}{dx} + c;$$

and, by integration with respect to x,
$$U = V + cx + c'.$$
Putting $c = 0$ and $c' = 0$, we have $U = V$, as was to be proved.

171. By the process of successive integration thus indicated we obtain the value of a function of two variables when its second derivative is given. The problem is, having an equation of the form
$$\frac{d^2u}{dxdy} = \phi(x, y), \qquad (2)$$
where $\phi(x, y)$ is supposed to be given, to find u, as a function of x and y. This we do by integrating first with respect to one of the variables, say x, which will give us the value of $\frac{du}{dy}$, because the first member of (2) is $D_x\frac{du}{dy}$. Then we integrate with respect to y, and thus get u.

As an example, let us take the equation
$$\frac{d^2u}{dxdy} = xy^2, \quad \text{or} \quad d.\frac{du}{dy} = xy^2 dx.$$
Integrating with respect to x, we have
$$\frac{du}{dy} = \frac{1}{2}x^2y^2 + h, \qquad (3)$$
h being a quantity independent of x, which we have commonly called an arbitrary constant. But, in accordance with a principle already laid down (§ 118), this so-called constant may be any quantity independent of x, and therefore any function we please to take of y.

Next, integrating (3) with respect to y, and putting
$$Y \equiv \int h dy,$$
we find $\qquad u = \frac{1}{6}x^2y^3 + Y + X,$
in which X is any quantity independent of y, and so may be an arbitrary function of x. Moreover, since h is an entirely arbitrary function of y, so is Y itself.

The student should now prove this equation by differentiating with respect to x and y in succession.

172. *Triple and Multiple Integrals.* The principles just developed may be extended to the case of integrals involving three or more independent variables. The expression

$$\int \int \int \phi(x, y, z) dx dy dz$$

means the result obtained by integrating $\phi(x, y, z)$ with respect to x, then that result with respect to y, and finally that result with respect to z. The final result is called **a triple integral**.

If we call $F(x, y, z)$ the final integral to be obtained, we have,

$$\frac{d^3 F(x, y, z)}{dx dy dz} = \phi(x, y, z);$$

and the problem is to find $F(x, y, z)$ from this equation when $\phi(x, y, z)$ is given.

Now, I say that to any integral obtained from this equation we may add, as arbitrary constants, three quantities: the one an arbitrary function of y and z; the second an arbitrary function of z and x; the third an arbitrary function of x and y. For, let us represent any three such functions by the symbols

$$[y, z], \quad [z, x], \quad [x, y],$$

and let us find the third derivative of

$$F(x, y, z) + [y, z] + [z, x] + [x, y] \equiv u$$

with respect to x, y and z. Differentiating with respect to x, y and z in succession, we obtain

$$\frac{du}{dx} = \frac{dF(x, y, z)}{dx} + \frac{d[z, x]}{dx} + \frac{d[x, y]}{dx};$$

$$\frac{d^2 u}{dx dy} = \frac{d^2 F(x, y, z)}{dx dy} + \frac{d^2 [x, y]}{dx dy};$$

$$\frac{d^3 u}{dx dy dz} = \frac{d^3 F(x, y, z)}{dx dy dz};$$

an equation from which the three arbitrary functions have

It is to be remarked that one or both of the variables may disappear from any of these arbitrary functions without changing their character. The arbitrary function of y and z, being any quantity whatever that does not contain x, may or may not contain y or z, and so with the others.

As an example, let it be required to find

$$u = \int\int\int (x-a)(y-b)(z-c)dxdydz.$$

Integrating with respect to z, and omitting the arbitrary function, we have

$$\int\int \tfrac{1}{2}(x-a)(y-b)(z-c)^2 dxdy.$$

Then integrating with respect to y,

$$\frac{du}{dx} = \int \tfrac{1}{4}(x-a)(y-b)^2(z-c)^2;$$

which gives, by integrating with respect to x, and adding the arbitrary functions,

$$u = \tfrac{1}{8}(x-a)^2(y-b)^2(z-c)^2 + [y, z] + [z, x] + [x, y].$$

The same principle may be extended to integrals with respect to any number of variables, or to **multiple integrals**.

The method may also be applied to the determination of a function of a single variable when the derivative of the function of any order is given.

EXERCISES.

1. $\int\int \dfrac{x}{y^2} dxdy.$ 2. $\int\int (x-a)(y-b)^2 dxdy.$

3. $\int\int\int xy^2 z^3 dxdydz.$ 4. $\int\int\int \dfrac{xy^2}{z^2} dxdydz.$

5. $\int\int\int (x-a)^2(y-b)(z-c)^2 dxdydz.$

6. $\int\int (x-a)^2 dx^2.$ 7. $\int\int\int (z+h)^2 dz^3.$

Ans. (6). $\tfrac{1}{12}(x-a)^4 + Cx + C'$, C and C' being arbitrary constants.

173. *Definite Double Integrals.* Let U be any function of x and y. By integration with respect to x, supposing y constant, we may form a definite integral

$$\int_{x_0}^{x_1} U dx \equiv U'.$$

From what has been shown in § 163, Rem., U' will be a function of y, x_0 and x_1. We may therefore form a second definite integral by integrating $U' dy$ between two limits y_0 and y_1. Thus we find an expression

$$\int_{y_0}^{y_1} U' dy = \int_{y_0}^{y_1} \int_{x_0}^{x_1} U dx dy,$$

which is a *definite double integral.*

The limits x_0 and x_1 of the first integration may be constants, or they may be functions of y.

If they are constants, the two integrations will be interchangeable, as shown for indefinite double integrals.

If they are functions of y they are not interchangeable, unless we make suitable changes in the limits.

174. *Definite Triple and Multiple Integrals.* A definite integral of any order may be formed on the plan just described. For example, in the definite triple integral

$$\int_{z_0}^{z_1} \int_{y_0}^{y_1} \int_{x_0}^{x_1} \phi(x, y, z) dx dy dz$$

the limits x_0 and x_1 of the first integration may be functions of y and z; while y_0 and y_1 may be functions of z. But z_0 and z_1 will be constants.

So, in any multiple integral, the limits of the first integration may be constants, or they may be functions of any or all the other variables. And each succeeding pair of limits may be functions of the variable which still remain, but cannot be functions of those with respect to which we have already integrated.

EXAMPLES AND EXERCISES.

1. Find the values of

$$\int_0^b \int_0^a xy^2 dx dy \quad \text{and} \quad \int_0^b \int_y^{2y} xy^2 dx dy.$$

It will be seen that in the first form the limits of x are constants, and in the second, functions of y.

First integrating with respect to x, we have for the indefinite integral

$$\int xy^2 dx = \tfrac{1}{2} x^2 y^2,$$

and for the two definite integrals

$$\int_0^a xy^2 dx = \tfrac{1}{2} a^2 y^2,$$

$$\int_y^{2y} xy^2 dx = \tfrac{3}{2} y^4.$$

Then, integrating these two functions with respect to y, we have

$$\int_0^b \int_0^a xy^2 dx dy = \tfrac{1}{2} \int_0^b a^2 y^2 dy = \tfrac{1}{6} a^2 b^3;$$

$$\int_0^b \int_y^{2y} xy^2 dx dy = \tfrac{3}{2} \int_0^b y^4 dy = \tfrac{3}{10} b^5.$$

Let us now see the effect of reversing the order of the integrations. First integrating with respect to y, we have

$$\int_0^b xy^2 dy = \tfrac{1}{3} xb^3 \equiv U.$$

Then integrating with respect to x, we have

$$\int U dx = \int_0^a \int_0^b xy^2 dy dx = \tfrac{1}{6} a^2 b^3,$$

the same result as when we integrated in the reverse order between the same constant limits.

2. Deduce $\int_0^{\frac{\pi}{2}} \int_0^{\pi} \cos(x+y) dx dy = -2.$

3. Deduce $\int_0^{\frac{1}{2}\pi} \int_0^{\frac{1}{2}\pi} \cos(x-y)dxdy = +4$.

4. Deduce $\int_0^b \int_0^a (x-a)(y-b)dxdy = \frac{1}{4}a^2b^2$.

5. Deduce $\int_0^{2a} \int_0^{2b} (x-a)(y-b)dxdy = \frac{1}{4}(2ab-a^2)(2ab-b^2)$.

6. Deduce $\int_b^{2a} \int_{-y}^{+y} (x-a)(y-b)dxdy = a^3b - \frac{1}{3}ab^3 - \frac{2}{3}a^4$.

175. *Product of Two Definite Integrals.*

THEOREM. *The product of the two definite integrals $\int_{x_0}^{x_1} X dx$ and $\int_{y_0}^{y_1} Y dy$ is equal to the double integral $\int_{y_0}^{y_1} \int_{x_0}^{x_1} XY dxdy$, provided that neither integral contains the variable of the other.*

For, by hypothesis, the integral $\int_{x_0}^{x_1} X dx \equiv U$ does not contain y. Therefore

$$U \int_{y_0}^{y_1} Y dy = \int_{y_0}^{y_1} UY dy = \int_{y_0}^{y_1} \int_{x_0}^{x_1} XY dxdy,$$

as was to be proved.

176. *The Definite Integral $\int_{-\infty}^{+\infty} e^{-x^2} dx$.* This integral, which we have already mentioned, is a fundamental one in the method of least squares, and may be obtained by the application of the preceding theorem. Let us put

$$k = \int_{-\infty}^{+\infty} e^{-x^2} dx = 2 \int_0^{+\infty} e^{-x^2} dx = 2 \int_0^{+\infty} e^{-y^2} dy. \,(\S 164)$$

Then, by the theorem,

$$k^2 = 4 \int_0^{+\infty} e^{-x^2} dx \int_0^{+\infty} e^{-y^2} dy = 4 \int_0^{+\infty} \int_0^{+\infty} e^{-(x^2+y^2)} dxdy.$$

Let us now substitute for y a new variable t, determined by the condition

$$y = tx.$$

Since, in integrating with respect to y, we suppose x constant, we must now put

$$dy = xdt.$$

Also, since t is infinite when y is infinite, and zero when y is zero, the limits of integration for t are also zero and infinity. Thus we have

$$k^2 = 4\int_0^{+\infty}\int_0^{+\infty} e^{-(1+t^2)x^2} x\, dx\, dt.$$

Since the limits are constants, the order of integration is indifferent. Let us then first integrate with respect to x. Since

$$xdx = \tfrac{1}{2}d\cdot x^2 = \frac{1}{2(1+t^2)} d\cdot(1+t^2)x^2,$$

the integral with respect to x is

$$\frac{1}{2(1+t^2)}\int_0^{+\infty} e^{-(1+t^2)x^2} d\cdot(1+t^2)x^2 = \frac{1}{2(1+t^2)}.$$

Then, integrating with respect to t,

$$k^2 = 2\int_0^{\infty} \frac{dt}{1+t^2} = \pi.$$

Hence

$$\int_{-\infty}^{+\infty} e^{-x^2} dx = \sqrt{\pi}.$$

CHAPTER VIII.

RECTIFICATION AND QUADRATURE.

177. *The Rectification of Curves.* In the older geometry to *rectify* a curve meant to find a straight line equal to it in length. In modern geometry it means to find an algebraic expression for any part of its length.

Let us put s for the length of the curve from an arbitrary fixed point C to a variable point P. If P' be another position of the variable point, we shall then have

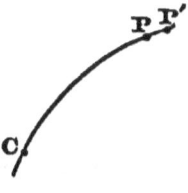

Fig. 50.

$$\varDelta s = PP'.$$

If PP' becomes infinitesimal, it has already been shown (§ 79) that we have, in rectangular co-ordinates,

$$ds = \sqrt{dx^2 + dy^2} = \sqrt{1 + \left(\frac{dy}{dx}\right)^2}\, dx = \sqrt{1 + \left(\frac{dx}{dy}\right)^2}\, dy, \quad (1)$$

and, in polar co-ordinates,

$$ds = \sqrt{r^2 + \left(\frac{dr}{d\theta}\right)^2}\, d\theta. \quad (2)$$

If both co-ordinates, x and y, are expressed in terms of a third variable u, we have

$$ds^2 = \left(\frac{dx}{du}du\right)^2 + \left(\frac{dy}{du}du\right)^2;$$

$$ds = \sqrt{\left(\frac{dx}{du}\right)^2 + \left(\frac{dy}{du}\right)^2}\, du.$$

The length of any part of the curve is then expressed by

the integral of any of these expressions taken between the proper limits. Thus we have

$$\left.\begin{aligned} s &= \int \left\{1 + \left(\frac{dy}{dx}\right)^2\right\}^{\frac{1}{2}} dx; \\ s &= \int \left\{1 + \left(\frac{dx}{dy}\right)^2\right\}^{\frac{1}{2}} dy; \\ s &= \int \left\{r^2 + \left(\frac{dr}{d\theta}\right)^2\right\}^{\frac{1}{2}} d\theta; \\ \text{or} \quad s &= \int \left\{\left(\frac{dx}{du}\right)^2 + \left(\frac{dy}{du}\right)^2\right\}^{\frac{1}{2}} du. \end{aligned}\right\} \quad (3)$$

In order to effect the integration it is necessary that the second members of (3) shall be so reduced as to contain no other variable than that whose differential is written; that is, we must have

$$ds = f(x)dx; \quad f(y)dy; \quad f(\theta)d\theta; \quad \text{or} \quad f(u)du.$$

Then we take for the limits of integration the values of x, y, θ or u, which correspond to the ends of the curve.

178. *Rectification of the Parabola.* From the equation of the parabola

$$y^2 = 2px$$

we derive $\quad ydy = pdx.$

We shall have the simplest integration by taking y as the independent variable. We then have

$$ds = \left\{1 + \left(\frac{dx}{dy}\right)^2\right\}^{\frac{1}{2}} dy; \quad pds = \{p^2 + y^2\}^{\frac{1}{2}} dy. \quad (a)$$

The formula (C) of § 145 gives

$$\int (p^2 + y^2)^{\frac{1}{2}} dy = \tfrac{1}{2} y (p^2 + y^2)^{\frac{1}{2}} + \tfrac{1}{2} p^2 \int \frac{dy}{(p^2 + y^2)^{\frac{1}{2}}}.$$

The method of § 132 gives

$$\int \frac{dy}{(p^2 + y^2)^{\frac{1}{2}}} = h - \log((p^2 + y^2)^{\frac{1}{2}} - y)$$
$$= h - \log p + \log((p^2 + y^2)^{\frac{1}{2}} + y).$$

Thus, putting $h' \equiv \frac{1}{2}p(h - \log p)$, the indefinite integral of (a) is

$$s = h' + \frac{1}{2}\frac{y}{p}(p^2 + y^2)^{\frac{1}{2}} + \frac{1}{2}p \log((p^2 + y^2)^{\frac{1}{2}} + y).$$

The arbitrary constant h' must be so taken that s shall vanish at the initial point of the parabolic arc. If we take the vertex as this point, we must have $s = 0$ for $y = 0$. Then

$$h' = -\tfrac{1}{2}p \log p.$$

We therefore have, for the length of a parabolic arc from the vertex to the point whose ordinate is y,

$$s = \frac{1}{2}\frac{y}{p}(p^2 + y^2)^{\frac{1}{2}} + \tfrac{1}{2}p \log \frac{(p^2 + y^2)^{\frac{1}{2}} + y}{p}. \qquad (4)$$

179. *Rectification of the Ellipse.* The formulæ for rectifying the ellipse take the simplest form when we express the co-ordinates in terms of the eccentric angle u; then (Analyt. Geom.)

$$x = a \cos u; \quad y = b \sin u.$$

We then have

$$dx = -a \sin u\, du; \quad dy = b \cos u\, du.$$

Then if e is the eccentricity, so that $a^2 e^2 = a^2 - b^2$,

$$ds = (a^2 \sin^2 u + b^2 \cos^2 u)^{\frac{1}{2}} du = a(1 - e^2 \cos^2 u)^{\frac{1}{2}} du;$$

$$s = a \int (1 - e^2 \cos^2 u)^{\frac{1}{2}} du.$$

This expression can be reduced to an *elliptic integral*: a kind of function which belongs to a more advanced stage of the calculus than that on which we are now engaged.

It may, however, be approximately integrated by development in series. We have, by the binomial theorem,

$$(1 - e^2 \cos^2 u)^{\frac{1}{2}} = 1 - \frac{1}{2}e^2 \cos^2 u - \frac{1 \cdot 1}{2 \cdot 4}e^4 \cos^4 u$$
$$- \frac{1 \cdot 1 \cdot 3}{2 \cdot 4 \cdot 6}e^6 \cos^6 u - \text{etc.}$$

The terms in the second member may be separately integrated by the formulæ (6), § 149, by putting $m = 0$ and $n = 2, 4, 6$, etc. We thus find

$$2\int \cos^2 u\, du = \sin u \cos u + u;$$
$$4\int \cos^4 u\, du = \sin u(\cos^3 u + \tfrac{3}{2}\cos u) + \tfrac{3}{2}u;$$
etc.   etc.   etc.

Since at one end of the major axis we have $u = 0$ and at the other end $u = \pi$, we find the length of one half of the ellipse by integrating between the limits 0 and π. Since $\sin u$ vanishes at both limits, we have

$$\int_0^\pi \cos^2 u\, du = \frac{1}{2}\pi;$$
$$\int_0^\pi \cos^4 u\, du = \frac{1\cdot 3}{2\cdot 4}\pi;$$
$$\int_0^\pi \cos^6 u = \frac{1\cdot 3\cdot 5}{2\cdot 4\cdot 6}\pi.$$

We thus find by substitution that the semi-circumference of the ellipse may be developed in powers of the eccentricity with the result

$$s = a\pi\left(1 - \frac{1}{2^2}e^2 - \frac{3}{2^2\cdot 4^2}e^4 - \frac{3^2\cdot 5}{2^2\cdot 4^2\cdot 6^2}e^6 - \cdots\right).$$

180. *The Cycloid.* The co-ordinates x and y of the cycloid are expressed in terms of the angle u through which the generating circle has moved by the equations (§ 80)

$$x = a(u - \sin u);$$
$$y = a(1 - \cos u).$$

Hence
$$ds^2 = dx^2 + dy^2 = a^2\{(1 - \cos u)^2 + \sin^2 u\}du^2$$
$$= 2a^2(1 - \cos u)du^2 = 4a^2 \sin^2 \tfrac{1}{2}u\cdot du^2.$$

By extracting the root and integrating,
$$s = h - 4a \cos \tfrac{1}{2}u.$$

If we measure the arc generated from the point where it meets the axis of abscissas, that is, where $u = 0$, we must have $s = 0$ for $u = 0$. This gives

$$h = 4a$$

and $$s = 4a(1 - \cos \tfrac{1}{2}u) = 8a \sin^2 \tfrac{1}{4}u.$$

This gives, for the entire length of the arc generated by one revolution of the generating circle,

$$s = 8a;$$

that is, four times the diameter of the generating circle.

181. *The Archimedean Spiral.* From the polar equation of this spiral (§ 82) we find

$$dr = a d\theta.$$

Hence $$ds = a(1 + \theta^2)^{\frac{1}{2}} d\theta.$$

Then the indefinite integral is (§ 147, Ex. 1)

$$s = \frac{a}{2} \left\{ \theta(1 + \theta^2)^{\frac{1}{2}} + \log C(\theta + (1 + \theta^2))^{\frac{1}{2}} \right\}.$$

If we measure from the origin we must determine the value of C by the condition that when $\theta = 0$, then $s = 0$. This gives $\log C = 0$; $\therefore C = 1$.

If instead of θ we express the length in terms of r, the radius vector of the terminal point of the arc, we shall have

$$s = \frac{1}{2} \frac{r}{a} (a^2 + r^2)^{\frac{1}{2}} + \frac{a}{2} \log \frac{(a^2 + r^2)^{\frac{1}{2}} + r}{a}.$$

182. *The Logarithmic Spiral.* The equation of this spiral (§ 83) gives

$$\frac{dr}{d\theta} = ale^{l\theta} = lr.$$

Hence $$ds = (1 + l^2)^{\frac{1}{2}} r d\theta.$$

To integrate this differential with respect to θ we should first substitute for r its value in terms of θ. But it will be better to adopt the inverse course, and express $d\theta$ in terms of dr. We thus have

$$ds = \frac{(1+l^2)^{\frac{1}{2}}}{l}dr;$$

whence
$$s = \frac{(1+l^2)^{\frac{1}{2}}}{l}r + s_0,$$

s_0 being the value of s for the pole.

If we put γ for the constant angle between the radius vector and the tangent, then (§§ 90–92) $l = \cot \gamma$, and we have

$$s = r \sec \gamma + s_0.$$

Between any two points of the curve whose radii-vectors are r_0 and r_1 we have

$$s = (r_1 - r_0) \sec \gamma.$$

Hence *the length of an arc of the logarithmic spiral is proportional to the difference between the radii-vectors of the extremities of the arc.*

EXERCISE.

1. Show that the differential of the arc of the lemniscate is

$$ds = \frac{a d\theta}{\sqrt{1 - 2 \sin^2 \theta}}.$$

(This expression can be integrated only by elliptic functions.)

183. *The Quadrature of Plane Figures.* In geometrical construction, to square a figure means to find a square equal to it in area. The operation of squaring is called *quadrature*.

In analysis, quadrature means the formation of an algebraic expression for the area of a surface.

In order to determine an area algebraically, the equation of the curve which bounds it must be given. Moreover, in order that the area may be completely determined by the bounding line, the latter must be a closed curve.

Then whatever the form of this curve, every straight line

QUADRATURE OF PLANE FIGURES.

must intersect it an even number of times. The simplest case is that in which a line parallel to the axis of Y cuts the boundary in two points. Then for every value of x the equation of the curve will give two values of y corresponding to ordinates terminating at P and Q. Let these values be y_0 and y_1.

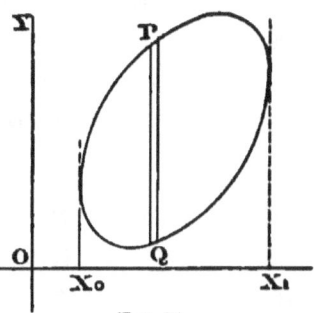

Fig. 51.

Then, the infinitesimal area included between two ordinates infinitely near each other will be

$$(y_1 - y_0)dx \equiv d\sigma.$$

The area given by integrating this expression will be

$$\sigma = \int_{x_0}^{x_1}(y_1 - y_0)dx,$$

in which the limits of integration are the extreme values of x corresponding to the points X_0 and X_1, outside of which the ordinate ceases to cut the curve.

The same principle may be applied by taking $(x_1 - x_0)dy$ as the element of the area. We then have

$$\sigma = \int_{y_0}^{y_1}(x_1 - x_0)dy.$$

If the curve is referred to polar co-ordinates, let S and T be two neighboring points of the curve, and let us put

$r \equiv OS;$
$r' \equiv OT;$
$\Delta\theta \equiv$ angle $SOT.$

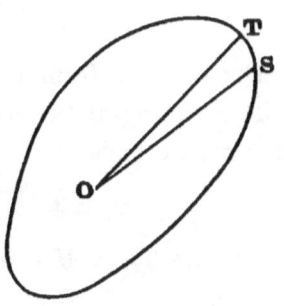

Fig. 52.

If we draw a chord from S to T, the area included between this chord and the curve will be of the third order (§ 78). The area of the triangle formed by this chord and the radii vectors will be $\frac{1}{2}rr'\sin\Delta\theta$. Now let $\Delta\theta$ become infinitesimal. OS will then approach r as its limit; the ratio of $\sin\Delta\theta$ to $\Delta\theta$ itself will approach unity, and the area of the triangle will approach that of the sector. Thus we shall have, for the differential of area,

$$d\sigma = \tfrac{1}{2}r^2 d\theta.$$

If the pole is within the area enclosed by the curve, the total area will be found by integrating this expression between the limits 0° and 360°. Thus we have, for the total area,

$$\sigma = \tfrac{1}{2}\int_0^{2\pi} r^2 d\theta.$$

184. *The Parabola.* As the parabola is not itself a closed curve, it bounds no area. But we may find the area of any segment cut off by a double ordinate MN. The equation of the curve gives, for the two values of y,

$$y_1 = +\sqrt{2px}; \quad y_0 = -\sqrt{2px}.$$

Hence

$$y_1 - y_0 = QP = \sqrt{8px};$$
$$d\sigma = \sqrt{8p}\cdot x^{\frac{1}{2}} dx.$$

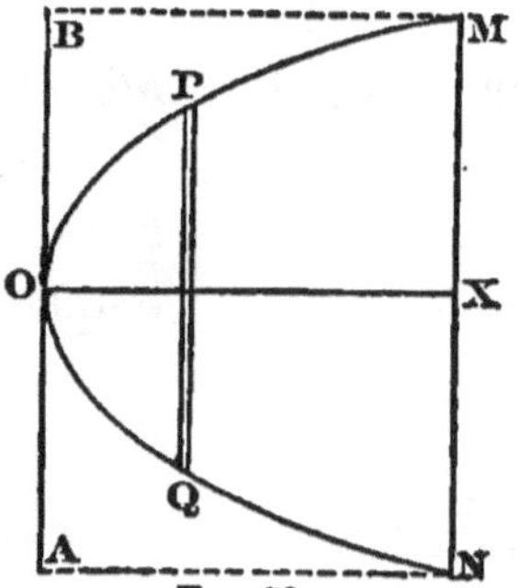

The indefinite integral is

$$\sigma = \tfrac{2}{3}\sqrt{2px^3} + C.$$

Fig. 53

For the area from the vertex to MN we put $x_1 \equiv OX$, and take the integral between the limits 0 and x_1. Calling this area σ_1, we have

$$\sigma_1 = \tfrac{2}{3}\sqrt{2px_1}\cdot x_1 = \tfrac{2}{3}x_1 y_1 = \tfrac{1}{3}x_1 \times 2y_1.$$

Because $2y_1 = MN$, it follows that the area $ABMN = 2x_1 y_1$. Hence:

THEOREM. *The area of a parabolic segment is two thirds that of its circumscribed rectangle.*

185. *The Circle and the Ellipse.* Referring the circle of radius a to the centre as the origin, the values of y will be

$$y = \pm (a^2 - x^2)^{\frac{1}{2}}.$$

Hence

$$\int (y_1 - y_0) dx = 2 \int (a^2 - x^2)^{\frac{1}{2}} dx$$

$$= x(a^2 - x^2)^{\frac{1}{2}} + a^2 \sin^{(-1)} \frac{x}{a} + h.$$

This expression, taken between appropriate limits, will give the area of any portion of the circle contained between two ordinates.

Taking the integral between the limits $-a$ and $+a$ gives, for the area of the circle,

$$\sigma = a^2 \sin^{(-1)} (+1) - a^2 \sin^{(-1)} (-1) = \pi a^2.$$

The Ellipse. From the equation of the ellipse referred to its centre and axes, namely,

$$\frac{x^2}{a^2} + \frac{y^2}{b^2} = 1,$$

we find
$$y = \pm \frac{b}{a} \sqrt{a^2 - x^2}.$$

The entire area will be

$$\int_{-a}^{+a} (y_1 - y_0) dx = 2 \frac{b}{a} \int_{-a}^{+a} (a^2 - x^2)^{\frac{1}{2}} dx = \pi ab.$$

The last integration is performed exactly as in the case of the circle.

186. *The Hyperbola.* Since the hyperbola is not a closed curve, it does not by itself enclose any area. But we may consider any area enclosed by an hyperbola and straight lines.

Let us first consider the area APM contained between the curve, the ordinate MP, and the segment AM of the major axis. The equation of the hyperbola referred to its centre and axes gives, for the value of y in terms of x,

$$y = \frac{b}{a} \sqrt{x^2 - a^2}.$$

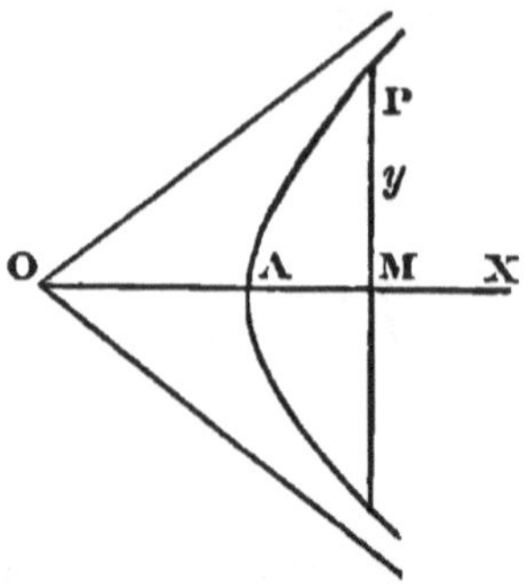

If we put x_1 for the value of the abscissa OM, then, since $OA = a$, the area AMP will be equal to the integral

$$\frac{b}{a} \int_a^{x_1} (x^2 - a^2)^{\frac{1}{2}} dx;$$

FIG. 54.

$$\int (x^2 - a^2)^{\frac{1}{2}} dx = \frac{1}{2} x(x^2 - a^2)^{\frac{1}{2}} - \frac{a^2}{2} \log \left[\frac{x}{a} + \left(\frac{x^2}{a^2} - 1\right)^{\frac{1}{2}} \right];$$

and for the definite integral between the limits a and x,

$$\text{Area } APM = \frac{1}{2} \frac{bx}{a}(x^2 - a^2)^{\frac{1}{2}} - \frac{ab}{2} \log \left[\frac{x}{a} + \left(\frac{x^2}{a^2} - 1\right)^{\frac{1}{2}} \right]$$
$$= \frac{1}{2} xy - \frac{ab}{2} \log \left[\frac{x}{a} + \left(\frac{x^2}{a^2} - 1\right)^{\frac{1}{2}} \right].$$

Now, $\frac{1}{2}xy$ is the area of the triangle OPM; we therefore conclude that the second term of the expression is the area included between OA, OP and the hyperbolic arc AP.

Much simpler is the area included between the curve, one asymptote, and two parallels to the other asymptote. The equation of the hyperbola referred to its asymptotes as axes of co-ordinates (which axes are oblique unless the hyperbola is equilateral) may be reduced to the form

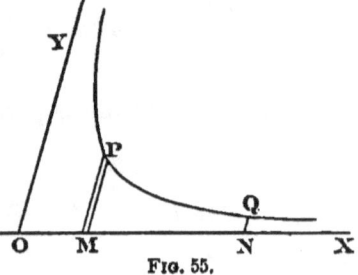

FIG. 55.

$$xy = \frac{ab}{2 \sin \alpha},$$

α being the angle between the axes. We readily see that the differential of the area is $ydx \times \sin \alpha$ instead of ydx simply. Hence for the area we have

$$\int y \sin \alpha dx = \int \frac{ab}{2x} dx = \frac{ab}{2} \log cx.$$

If we take the area between the limits $OM \equiv x_0$ and $OM \equiv x_1$, the result will be

$$\int_{x_0}^{x_1} \frac{ab}{2x} dx = \frac{ab}{2} \log \frac{x_1}{x_0}.$$

We note that this area becomes infinite when x_0 becomes zero or when x_1 becomes infinite, showing that the entire area is infinite.

187. *The Lemniscate.* The equation of this curve in polar co-ordinates is (§ 81)

$$r^2 = a^2 \cos 2\theta.$$

It will be noted that r becomes imaginary when θ is contained between 45° and 135°, or between 225° and 315°.

The integral expression for the area is

$$\tfrac{1}{2} \int r^2 d\theta = \tfrac{1}{2} a^2 \int \cos 2\theta d\theta = \tfrac{1}{4} a^2 \sin 2\theta.$$

To find the area of the right-hand loop of the curve we must take this integral between the limits $\theta = -45°$ and $\theta = +45°$, for which $\sin 2\theta = -1$ and $+1$. Hence

Half area $= \tfrac{1}{2} a^2$;
Total area $= a^2$.

Hence *the area of each loop of the lemniscate is half the square on the semi-axis.*

188. *The Cycloid.* By differentiating the expression for the abscissa of a point of the cycloid we have

$$dx = a(1 - \cos u)du.$$

Hence

$$\int_0^{2\pi} y\,dx = a^2 \int_0^{2\pi}(1-\cos u)^2 du = a^2 \int_0^{2\pi}(\tfrac{3}{2}-2\cos u + \tfrac{1}{2}\cos 2u)du.$$

The indefinite integral is

$$\tfrac{3}{2}u - 2\sin u + \tfrac{1}{4}\sin 2v.$$

To find the whole area we take the definite integral between the limits 0 and 2π. Thus we find

Area of cycloid $= 3\pi a^2$,

or three times the area of the generating circle.

EXERCISES.

1. Show that the theorem of § 184 is true only of the parabola.

To do this we must find what the equation of a curve must be in order that the theorem may be true. The theorem is

$$\int y\,dx = \tfrac{2}{3}xy.$$

Differentiating both members, we have

$$y\,dx = \tfrac{2}{3}x\,dy + \tfrac{2}{3}y\,dx;$$

$$\therefore\ 2\frac{dy}{y} = \frac{dx}{x}.$$

Then, integrating both members,

$$\log y^2 = \log cx\ ;\ \ \therefore\ y^2 = cx,$$

c being an arbitrary constant. This is the equation of a parabola whose parameter is $\tfrac{1}{2}c$.

2. Show that the equation of a curve the ratio of whose area to that of the circumscribed rectangle is $m:n$ must be of the form

$$y^m = cx^{n-m}.$$

CHAPTER IX.

THE CUBATURE OF VOLUMES.

189. *General Formulæ for Cubature.* In the ancient Geometry to *cube* a solid meant to find the edge of a cube whose volume should be equal to that of the solid. In Analytic Geometry it means to find an expression for the volume of a solid.

Let us have a solid the bounding surface of which is defined by an equation between rectangular co-ordinates. Let the solid be cut by a plane PL parallel to the plane of YZ, and let u be the area of the plane section thus formed. If we now cut the solid by a second plane, parallel to PL and infinitely near it, that portion of the solid contained between

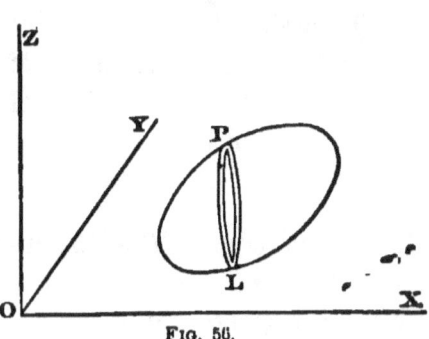

Fig. 56.

the planes will be a slice of area u and thickness dx, dx being the infinitesimal distance between the planes.

If, then, we put v for the volume of that part of the solid contained between any two planes parallel to YZ, we have

$$dv = u\,dx,$$

and $$v = \int_{x_0}^{x_1} u\,dx, \qquad (1)$$

x_0 and x_1 being the distances of the cutting planes from the origin O.

If we take for x_0 and x_1 the extreme values of x for any part of the solid, the above expression will give the total volume of the solid.

In order to integrate (1), we must express u as a function of x. That is, we must find a general expression in terms of x for the area of any section of the solid by a plane parallel to that of XY. This is to be done by the equation of the bounding surface of the solid.

Of course we may form the infinitesimal slices by planes perpendicular to the axis of Y or of Z as well as of X.

190. *The Sphere.* The equation of a sphere referred to its centre as the origin is

$$x^2 + y^2 + z^2 = a^2.$$

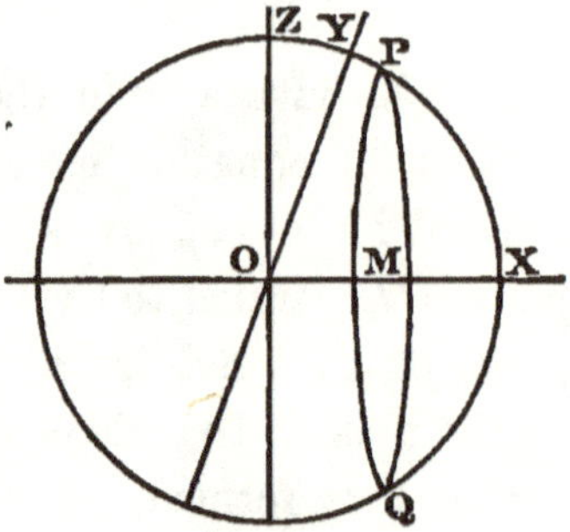

If we cut the sphere by a plane PMQ parallel to the plane of YZ, and having the abscissa $OM = x$, the equation of the circle of intersection will be

$$y^2 + z^2 = a^2 - x^2;$$

Fig. 57.

that is, the radius MP of the circle will be $\sqrt{a^2 - x^2}$, and its area will be $\pi(a^2 - x^2)$. Hence the differential of the volume of the sphere will be

$$dv = \pi(a^2 - x^2)dx,$$

and the indefinite integral will be

$$v = \pi(a^2 x - \tfrac{1}{3}x^3) + C.$$

The extreme limits of x for the sphere are

$$x_0 = -a \quad \text{and} \quad x_1 = +a.$$

Taking the integral between these limits, we have

$$\text{Volume of sphere} = \tfrac{4}{3}\pi a^3.$$

CUBATURE OF VOLUMES.

191. Volume of Pyramid. Let the pyramid be placed with its vertex at the origin, and its base parallel to the plane of XY. Let us also put $h = OZ$ its altitude; a, the area of its base. Let it be cut by a plane $EFGH$ parallel to its base.

It is shown in Geometry that the section $EFGH$ is similar to the base, and that

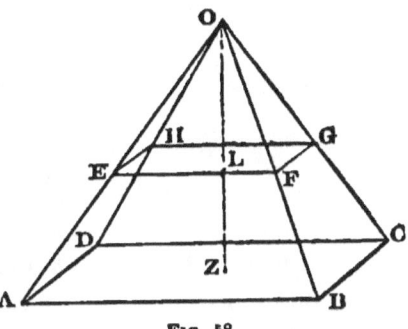

Fig. 58.

the ratio of any two homologous sides, as EF and AB, is the same as the ratio $OL : OZ$. Because the areas of polygons are proportional to the squares of their homologous sides,

$$\therefore \text{Area } EFGH : \text{Area } ABCD = OL^2 : OZ^2.$$

Putting Area $ABCD = a$, $OL = z$ and $OZ = h$,

$$\text{Area } EFGH = \frac{az^2}{h^2}.$$

The volume of the pyramid is therefore

$$V = \int_0^h \frac{az^2 dz}{h^2} = \frac{1}{3} ah.$$

That is, one third the altitude into the base.

The same formulæ apply to the cone.

192. The Ellipsoid. The equation of the ellipsoid referred to its centre and axes is

$$\frac{x^2}{a^2} + \frac{y^2}{b^2} + \frac{z^2}{c^2} = 1,$$

a, b and c being the principal semi-axes.

If we cut the ellipsoid by the plane whose equation is $x = x'$, the equation of the section will be

This is the equation of an ellipse whose semi-axes are
$\dfrac{b}{a}\sqrt{a^2 - x'^2}$ and $\dfrac{c}{a}\sqrt{a^2 - x'^2}$.

Hence its area is $\dfrac{\pi bc(a^2 - x'^2)}{a^2}$.

Then, by integration between the limits $-a$ and $+a$, we find
Volume of ellipsoid $= \tfrac{4}{3}\pi abc$.

From the known expression for the area of an ellipse (πab) it is readily found that the volume of an elliptic cylinder circumscribing any ellipsoid is $2\pi abc$. Hence we conclude:

The volume of an ellipsoid is two thirds that of any right elliptic cylinder circumscribed about it.

193. *Volume of any Solid of Revolution.* In order that a solid of revolution may have a well-defined volume it must be generated by the revolution of a curve or unbroken series of straight or curve lines terminating at two points, Q and R, of the axis of revolution.

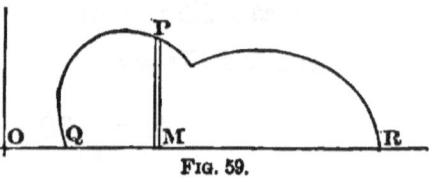

Fig. 59.

As an element of the volume we take two planes infinitely near each other and perpendicular to the axis of revolution. Every such plane cuts the solid in a circle. If we place the origin at O, take the axis of revolution as that of X, and let

$OM \equiv x$ be the abscissa of any point P of the curve, and
$MP \equiv y$ its ordinate,

then the section of the solid through M will be a circle of radius y, whose area will therefore be πy^2.

Hence the volume contained between two planes at distance dx will be
$$\pi y^2 dx,$$
and the volume between two sections whose abscissas are x_0 and x_1 will be
$$V = \int_{x_0}^{x_1} \pi y^2 dx. \tag{1}$$

CUBATURE OF VOLUMES.

If the two co-ordinates are expressed in terms of a third variable u by the equations

$$x = \phi(u), \quad y = \psi(u),$$

we have $\qquad dx = \phi'(u)du.$

Putting u_0 and u_1 for the values of u corresponding to x_0 and x_1, the expression (1) for the volume will become

$$V = \pi \int_{u_0}^{u_1} [\psi(u)]^2 \phi'(u) du. \qquad (2)$$

The equations (1) and (2) give the volume $AA'B'B$ generated by the revolution of any arc AB of the given curve, and of the ordinates MA and NB of the extremities of the arc. The limits of integration for x are $OM = x_0$ and $ON = x_1$. To

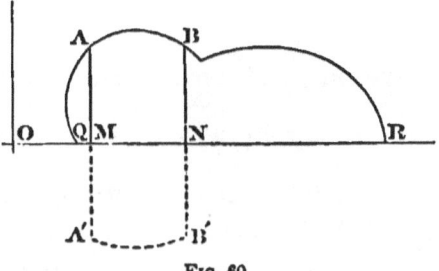

Fig. 60.

find the entire volume generated we must extend these limits to the points (if any) at which the curve intersects the axis of revolution.

194. *The Paraboloid of Revolution.* The equation of the parabola being $y^2 = 2px$, we readily find from (1) a result leading to the following theorem, which the student should prove for himself:

THEOREM. *The volume of a paraboloid of revolution is one half that of the circumscribed cylinder.*

195. *The Volume Generated by the Revolution of a Cycloid around its Base.* From the equations of the cycloid in terms of

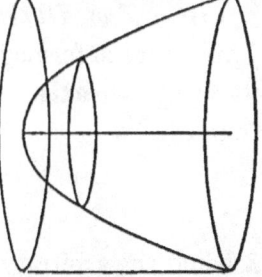

Fig. 61.

the angle through which the generating circle has moved, we find the element of the volume to be

$$dV = \pi a^3 (1 - \cos u)^3 du.$$

Hence

$$V = \pi a^3 \int (1 - 3 \cos u + 3 \cos^2 u - \cos^3 u) du.$$

By the method of §§ 149, 150, with simple reductions, we find

$$\int \cos^2 u \, du = \tfrac{1}{2} u + \tfrac{1}{4} \sin 2u;$$

$$\int \cos^3 u \, du = \int (1 - \sin^2 u) d . \sin u = \sin u - \tfrac{1}{3} \sin^3 u$$
$$= \tfrac{3}{4} \sin u + \tfrac{1}{12} \sin 3u.$$

We thus find, for the indefinite integral,

$$V = \pi a^3 (\tfrac{5}{2} u - \tfrac{15}{4} \sin u + \tfrac{3}{4} \sin 2u - \tfrac{1}{12} \sin 3u).$$

The total volume formed by the revolution of one arc of the cycloid is found by taking the integral between the limits $u = 0$ and $u = 2\pi$. The volume thus becomes

$$V = 5 \pi^2 a^3,$$

from which follows the theorem:

The volume generated by the revolution of a cycloid around its base is five eighths that of the circumscribed cylinder.

196. *The Hyperboloid of Revolution of Two Nappes.* This figure is formed by the revolution of an hyperbola about its transverse axis. The general expression for the volume is found to be

$$V = \frac{\pi b^2}{3 a^2} (x^3 - 3 a^2 x + h),$$

h being the arbitrary constant of integration. If we consider that part of the infinite solid cut off by a plane perpendicular to the transverse axis, we must determine h by the condition that V shall vanish when $x = a$, because then the plane will be a tangent at the vertex of the hyperboloid, and the volume will become zero. This condition gives

$$h = 3a^2 - a^2 = 2a^2.$$

Thus we have

$$V = \frac{\pi b^2}{3a^2}(x^3 - 3a^2 x + 2a^3) = \frac{\pi b^2}{3a^2}(x - a)^2(x + 2a). \quad (1)$$

By the same revolution whereby the hyperbola describes an hyperboloid of revolution the asymptotes will describe a cone. Let us compare the volume just found for the hyperboloid with that of the asymptotic cone, cut off by the same plane which cuts off the hyperboloid. The equation of the generating asymptote being

$$ay = bx,$$

we find for the volume of the cone

$$V' = \frac{\pi}{3} y^2 x = \frac{\pi b^2}{3a^2} x^3. \quad (2)$$

The difference between (1) and (2) will be the volume of the cup-shaped solid formed by cutting the hyperboloid out of the cone. Calling this volume V'', we find

$$V'' = \pi b^2 (x - \tfrac{2}{3}a). \quad (3)$$

This is equal to the volume of a circular cylinder of which the diameter is the conjugate axis of the hyperbola, and the altitude $x - \tfrac{2}{3}a$.

This result is intimately associated with the following theorem, the proof of which is quite easy:

If a plane perpendicular to the axis of revolution cut an hyperbola of two nappes and its asymptotic cone, the area of the plane contained between the circular sections is constant and equal to the area of the circle whose diameter is the conjugate axis.

197. Ring-shaped Solids of Revolution. If any completely bounded plane figure $APQB$ revolve around an axis OX lying in its own plane, but wholly outside of it, it will describe a ring-shaped solid.

To investigate such a solid, let the ordinate MP cut the figure in the points Q and P, and let us put

$$y_1 \equiv MQ; \quad y_2 \equiv MP.$$

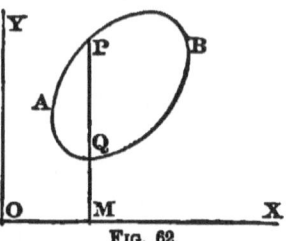

FIG. 62.

The points P and Q will describe two circles which will contain between them the sectional area

$$\pi(y_2^2 - y_1^2).$$

Taking two ordinates at the infinitesimal distance dx, the corresponding infinitesimal element of volume will be

$$dV = \pi(y_2^2 - y_1^2)dx. \tag{1}$$

The integral

$$V = \pi \int_{x_0}^{x_1} (y_2^2 - y_1^2)dx = \pi \int_{x_0}^{x_1} (y_2 + y_1)(y_2 - y_1)dx$$

will express the volume of that part of the solid contained between the two planes whose respective abscissas are x_0 and x_1. By taking for x_0 and x_1 the abscissas of the extreme points A and B, V will express the total volume of the solid.

198. Application to the Circular Ring. Let the figure AB be a circle of radius c, whose centre is at the distance b from the axis of revolution. Let us also put

$a \equiv$ the abscissa of the centre.

We then have

$$y_1 = b - \sqrt{c^2 - (x-a)^2};$$
$$y_2 = b + \sqrt{c^2 - (x-a)^2};$$
$$y_2 + y_1 = 2b;$$
$$y_2 - y_1 = 2\sqrt{c^2 - (x-a)^2};$$

$$V = 4\pi b \int_{x_0}^{x_1} [c^2 - (x-a)^2]^{\frac{1}{2}} dx.$$

The limits of integration for the whole volume are

$$x_0 = a - c \quad \text{and} \quad x_1 = a + c.$$

If we put

$$z \equiv x - a,$$

the total volume will become

$$V = 4\pi b \int_{-c}^{+c} (c^2 - z^2)^{\frac{1}{2}} dz.$$

By substituting the known value of the definite integral, we have

$$V = 2\pi^2 bc^2.$$

The area of the generating circle is πc^2, and the circumference of the circle described by its centre is $2\pi b$. The product of these two quantities is $2\pi^2 bc^2$. Hence:

The volume of a circular ring is equal to the product of the area of its cross-section into the circumference of its central circle.

EXAMPLES AND EXERCISES.

1. Compare the cycloid with the semi-ellipse having the same axes as the cycloid, and show the following relations between them:

α. The maximum radius of curvature of the ellipse (at the point B) is greater than that of the cycloid in the ratio $\pi^2 : 8$, or $5 : 4$, nearly.

β. The area of the semi-ellipse is greater than that of the cycloid in the ratio $\pi : 3$.

γ. The volume of the ellipsoid of revolution around the axis OX is greater than that generated by the revolution of the cycloid in the ratio $16 : 15$.

199. *Quadrature of Surfaces of Revolution.* Let us put

$\Delta s \equiv$ a small arc PQ of a curve revolving round an axis OX;
$y \equiv$ the distance of P from the axis OX;
$y' \equiv$ the distance of Q from the axis OX.

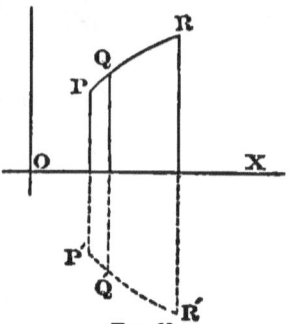

Fig. 63.

Considering Δs as a straight line, the surface generated by it will be the curved surface of the frustum of a cone. If we put

$\Delta \sigma \equiv$ the area of this curved surface, we have, by Geometry,

$$\Delta \sigma = \pi(y + y')\Delta s.$$

Now let Δs become infinitesimal. Then y' will approach y as its limit, and we shall have, for the differential of the surface,

$$d\sigma = 2\pi y \, ds = 2\pi y \left[1 + \left(\frac{dy}{dx}\right)^2 \right]^{\frac{1}{2}} dx.$$

This expression, when integrated between the limits x_0 and x_1, will give the area of that portion of the surface for which the co-ordinates x are contained between x_0 and x_1.

The modifications and transformations of this formula so as to apply it to cases when another axis than that of Y is the axis of revolution, or when the equation of the curve is not in the form $y = \phi(x)$, can be made by the student himself.

200. *Examples of Surfaces of Revolution.* The process of applying the general formula for $d\sigma$ to special cases is so like that already followed in quadrature and cubature that the briefest indications will suffice to guide the student.

Surface of the Sphere. Supposing the equation of the generating circle to be written in the form

$$x^2 + y^2 = a^2,$$

we shall find the differential of the surface to be
$$d\sigma = 2\pi a dx.$$

From this we may easily prove the following:

THEOREM I. *If a sphere be cut by any number of parallel and equidistant planes, the curved surfaces of the spherical zones contained between the planes will all be equal to each other.*

THEOREM II. *The total surface of a sphere is equal to the product of its diameter and circumference.*

Surface generated by the Revolution of a Cycloid. We shall find the differential of the surface to be
$$d\sigma = 8\pi a^2 \sin^3 \tfrac{1}{2}u\, du.$$

By a formula found in Trigonometry, we have
$$8 \sin^3 v = 6 \sin v - 2 \sin 3v.$$

Hence, putting $v \equiv \tfrac{1}{2}u$,
$$d\sigma = 4\pi a^2 (3 \sin v - \sin 3v)dv.$$

The whole surface is obtained by integrating between the limits $u = 0$ and $u = 2\pi$; that is, $v = 0$ and $v = \pi$. We thus find, for the total surface,
$$\sigma = \tfrac{64}{3}\pi a^2.$$

Hence the theorem:

The total surface generated by the revolution of a cycloid about its base is four thirds the surface of the greatest inscribed sphere.

The Paraboloid of Revolution. Taking the integral between the limits zero and x, we have for the curved surface
$$\sigma = \frac{2}{3}\frac{\pi}{p}\{(p^2 + 2px)^{\frac{3}{2}} - p^3\}.$$

THE END.

$\cos^{-1}x = u$
$1 = \sin u \cdot k$
$k \cdot v = \cos u \cdot k$
$ = \cos u$

$\dfrac{-1}{\sqrt{1-x^2}} \cdot \sin u = \cos^{-1}x \cdot$

$(1-x^2)\sin^{-1}x \cdot \cos^{-1}x + \dfrac{x}{2}\int \dfrac{dx}{\sqrt{1-x^2}} = (\ C\ )$

$ - \dfrac{(1-x^2)\sin^{-1}x}{2} + \sin^{-1}x$

$ - \dfrac{(1-x^2)}{2} + x$

$I = \dfrac{\sin^{-1}x \cos^{-1}x}{2} + x \cdot \sin^{-1}x$

www.ingramcontent.com/pod-product-compliance
Lightning Source LLC
Chambersburg PA
CBHW030753230426
43667CB00007B/953